Sport and Exercise Physiology

The INSTANT NOTES series

Series Editor: B.D. Hames, School of Biochemistry and Molecular Biology, University of Leeds, Leeds, UK

Animal Biology 2nd edition
Ecology 2nd edition
Genetics 2nd edition
Microbiology 2nd edition
Chemistry for Biologists 2nd edition
Immunology 2nd edition
Biochemistry 2nd edition
Molecular Biology 2nd edition
Neuroscience
Developmental Biology
Plant Biology
Bioinformatics
Sport and Exercise Physiology

Chemistry series
Consulting Editor: Howard Stanbury

Organic Chemistry 2nd edition
Inorganic Chemistry 2nd edition
Physical Chemistry
Medicinal Chemistry
Analytical Chemistry

Psychology series
Sub-series Editor: Hugh Wagner, Dept of Psychology, University of Central Lancashire, Preston, UK

Psychology
Cognitive Psychology
Physiological Psychology

Forthcoming titles
Sport and Exercise Psychology
Sport and Exercise Mechanics

Sport and Exercise Physiology

K. Birch

School of Sport and Exercise Sciences,
University of Leeds, Leeds, UK

D. MacLaren

The Research Institute for Sport and Exercise Sciences,
Liverpool John Moores University, Henry Cotton Campus,
15–21 Webster Street, Liverpool, UK

and

K. George

The Research Institute for Sport and Exercise Sciences,
Liverpool John Moores University, Henry Cotton Campus,
15–21 Webster Street, Liverpool, UK

BIOS Scientific Publishers
Taylor & Francis Group

© Garland Science/BIOS Scientific Publishers, 2005

First published 2005

A CIP catalogue record for this book is available from the British Library.

ISBN 1 85996 2491

Garland Science/BIOS Scientific Publishers
4 Park Square, Milton Park,
Abingdon, Oxon OX14 4RN, UK and

270 Madison Avenue, New York,
NY 10016, USA
World Wide Web home page: www.garlandscience.com

Garland Science/BIOS Scientific Publishers is a member of the Taylor & Francis Group

Distributed in the USA by
Fulfilment Center
Taylor & Francis
10650 Toebben Drive
Independence, KY 41051, USA
Toll Free Tel.: +1 800 634 7064; E-mail: taylorandfrancis@thomsonlearning.com

Distributed in Canada by
Taylor & Francis
74 Rolark Drive
Scarborough, Ontario M1R 4G2, Canada
Toll Free Tel.: +1 877 226 2237; E-mail: tal_fran@istar.ca

Distributed in the rest of the world by
Thomson Publishing Services
Cheriton House
North Way
Andover, Hampshire SP10 5BE, UK
Tel.: +44 (0)1264 332424; E-mail: salesorder.tandf@thomsonpublishingservices.co.uk

Library of Congress Cataloging-in-Publication Data

Birch, K. (Karen)
    Instant notes in sport and exercise physiology/K. Birch, D. MacLaren, K. George. — 1st ed.
        p. cm.
    ISBN 1-85996-249-1
    1. Exercise—Physiological aspects. 2. Sports—Physiological aspects. I. MacLaren, D. (Don)
II. George, K. (Keith) III. Title.

QP301.B4786 2004
612'.044—dc22
                                    2004014197

Production Editor: Catherine Jones
Typeset by Phoenix Photosetting, Chatham, Kent, UK
Printed by Biddles Ltd, Guildford, UK, www.biddles.co.uk

CONTENTS

ABBREVIATIONS

AACVPR	American Association of Cardiovascular and Pulmonary Rehabilitation
ACOG	American College of Obstetricians and Gynecologists
ADP	adenosine diphosphate
AHA	American Heart Association
ANS	autonomic nervous system
ATP	adenosine triphosphate
BMI	body mass index
cAMP	cyclic adenosine monophosphate
CHD	coronary heart disease
CK	creatine kinase
CNS	central nervous system
DEXA	dual-energy X-ray absorptiometry
DLW	doubly labeled water
DOMS	delayed-onset muscle soreness
ECG	electrocardiogram
$FEV_{1.0}$	forced expiratory volume in 1 second
FFM	fat-free mass
FG	fast glycolytic
FOG	fast oxidative-glycolytic
FVC	forced vital capacity
HCO_3^-	bicarbonate ion
HR	heart rate
IDDM	insulin-dependent diabetes mellitus
IGF-I	insulin-like growth factor I
LCT	long-chain triglyceride
LSD	long slow distance
MCT	medium-chain triglyceride
MET	metabolic equivalent
MLTPA	Minnesota Leisure Time Physical Activity Questionnaire
MRFIT	Multiple Risk Factor Intervention Trial
NIDDM	non-insulin-dependent diabetes mellitus
NIR	near infrared reactance
PAR-Q	Physical Activity Readiness Questionnaire
PASE	Physical Activity Scale for the Elderly
PCO_2	partial pressure of carbon dioxide
PO_2	partial pressure of oxygen
PCR	phosphocreatine/creatine phosphate
PDH	pyruvate dehydrogenase
PFK	phosphofructokinase
PNF	proprioceptive neuromuscular facilitation
PNS	peripheral nervous system
$\dot{Q}$	cardiac output
RER	respiratory exchange ratio
RM	repetition maximum
RMR	resting metabolic rate
ROM	range of motion
RPE	rating of perceived exertion
RQ	respiratory quotient
SCUBA	self-contained underwater breathing apparatus
SI	System International
SO	slow oxidative
SV	stroke volume
TEA	thermic effect of activity
TEF	thermic effect of food
TIA	transient ischemic attack
$\dot{V}CO_2$	volume of carbon dioxide produced
VE	volume expired
$\dot{V}E$	ventilatory volume
$\dot{V}O_2$	volume of oxygen consumed per minute
$\dot{V}O_{2max}$	maximum rate of oxygen consumed per minute

PREFACE

Sport and Exercise Science has become a hugely popular degree subject at university and college alike, and thankfully more and more health professionals utilize the scientific principles of exercise, training, sport and rehabilitation. With this in mind it seemed that a quick reference, or revision text in the physiology of exercise was sorely missing. We had the idea that a text of this nature would be applicable to undergraduate students studying Exercise Physiology as a major component of their degree, to students opting to study the elective as an option, and then of course to those professionals and interested parties who need a guide to the essentials of the subject. Hopefully this text will be used by all of these parties. We acknowledge that the title of the text is grammatically incorrect, and should in fact read the 'Physiology of Exercise.' After much debate and thought over this problem we decided to stay with 'Exercise Physiology,' mainly because it is the name given to most undergraduate modules of this nature. The Key Notes sections of this book should highlight the important revision areas within each topic of interest. The following sections provide the major detail of these topics, whilst also highlighting where topics interlink. We have attempted to keep this detail to the essential information, and to use easy diagrams to help underpin the knowledge. Students and interested parties alike should therefore become very familiar with the essential components of exercise physiology without having to read through huge tomes. We hope that once the essential information is grasped that the reader will be enthused to read further in what is (we think) a fascinating subject.

Karen Birch,
August 2004

A1 EXERCISE PHYSIOLOGY

Key Notes

Physical activity	Physical activity is defined as skeletal muscle contraction that results in an increase in energy expenditure. Movement may not always occur given that isometric contractions require energy but do not result in movement.
Exercise and sport	Exercise is defined as repetitive physical activity or muscular contraction aimed at improving or maintaining fitness or health. Sport is physical activity or movement involving rules and competition.
Exercise physiology	Exercise physiology is the scientific discipline involving the examination of how acute and chronic physical activity influences the structure and function of the human body.
Exercise science vs sport science	Exercise science is the study of how physical activity or exercise affects human health and function or vice versa. Sports science is the study of how physical activity or function affects sports performance or vice versa.
Ergometry	Ergometry refers to the methods and devices used to either control or measure work rate or intensity of exercise.
SI units	System International units are units of measurement used as a standardized method of reporting results.
Related topics	This chapter underpins the entire book!

Physical activity

Physical activity is defined as any skeletal muscle contraction that results in an increased energy expenditure. Thus physical activity might involve walking for a bus or training for a football match and therefore includes all exercise and sports activities. Isometric muscular contractions (those where muscle length does not change) require energy expenditure but do not necessarily result in movement occurring (for example holding the Maltese Cross in gymnastics, or pushing forwards in the rugby scrum).

Exercise and sport

Exercise is defined as repetitive physical activity or movement aimed at improving or maintaining fitness or health. Aerobics classes, weight training and working out in the gym are examples of exercise. **Sport** is physical activity or movement involving rules and competition. Football, hockey, basketball and badminton are examples of specific sporting activities.

Exercise physiology

Exercise physiology is the discipline involving the examination of how physical activity, exercise or sport influences the structure and function of the human body. The discipline is usually studied by assessing how movement affects the systems of the body (i.e. the cardiorespiratory system, the nervous system, the musculoskeletal system and the endocrine system), the cells of the body and subcellular molecules. In this way the exercise physiologist can study the immediate effects of movement on human function, or the long-term effects of regular physical activity. The immediate or **acute effects** of movement or physical activity might, for example, involve assessing what happens to the heart rate during 10 minutes of running at a set speed on a treadmill. The long-term, or **chronic (adaptive) effects** of physical activity on the other hand might involve assessing how resting heart rate is affected by 12 weeks of a set exercise or training program.

Exercise science vs sports science

Exercise physiology underpins both exercise and sports science. In the example above the exercise or training program may be for a group of individuals classed as obese, or for a hockey or football team. The chronic effects of the program upon physiological function might then either underpin optimal health and function, or optimal sports performance. This is the difference between exercise and sport science. **Exercise science** is the study of how physical activity or exercise affects human health and function or vice versa. **Sports science** is the study of how physical activity or function affects sports performance or vice versa.

Ergometry

Ergometry refers to the methods used to either control or measure work rate/exercise intensity, in other words the methods underpinning the majority of laboratory-based exercise physiology experiments. The devices used in these processes are called **ergometers** (typically the stepping bench, cycle ergometer (both electronically and friction braked), treadmill and arm ergometer). An understanding of how work is measured on any ergometer allows the exercise physiologist to estimate **energy expenditure**, and thus to plan training, rehabilitation or disease intervention sessions.

SI units

When working with ergometers the exercise physiologist must recognize that the correct standardized units of measurement are **System International Units** (SI units). However, many other more traditional units are used in the laboratory and thus the exercise physiologist needs to be familiar with converting one set of units to another (*Table 1, Section A2*). An example of this practice is that of referring to distance in meters, miles or kilometers. The distance is the same but the units of measurement differ. Students often become confused with units so it is important to remember that the measure is the same, but the units used to describe the measure may vary. More examples of these conversions are provided in Section A2.

A2 FORCE, WORK AND POWER

Key Notes

Overview	Understanding concepts such as force, work and power will allow the exercise physiologist to be able to undertake experiments in the laboratory. This is especially the case as these terms are used to describe movement conducted on ergometers.
Force	Force is that which changes the state of rest or motion of an object.
Work	Work is the quantification of force operating upon a mass causing it to change location.
Power	Power is defined as the rate at which work is performed.
Related topics	Work and power performed on the cycle ergometer and treadmill (A3)

Overview

The exercise physiologist spends a great deal of time in the laboratory measuring physiological responses to movement. The use of ergometers to undertake this task means that terms such as work, force and power need to be understood. Force may be applied to the cycle ergometer in order to provide a load against which to cycle. The subject will then turn the pedals producing work, and cycling at a specific work rate. The following sections hope to help clarify these terms.

Force

Force is that which acts to change the state of rest or motion of an object. Traditionally in the exercise physiology laboratory force was quantified in kiloponds (kp), where 1 kp represents the force exerted by 1 kg mass in normal gravity (i.e. 1 kg = 1 kp). The SI unit of force is the Newton (N) and 1 kp, or 1 kg = 9.80665 N.

Work

Work is a quantification of the force operating upon a mass that causes the mass to change its physical location. In this manner work can also be defined as:

$$\text{work} = \text{force} \times \text{distance}$$

Consequently, movement of a 2 kg mass over a 1 meter distance would require the following work to be performed:

Note: kg is a unit of mass and not force
Force is correctly quantified in Newtons (N: i.e. the correct units to use)
Thus: 1 kg = 9.81 N (see *Table 1*)
2 kg = 9.81 × 2 = 19.62 N

Consequently, work can also be quantified using Newtons as a measure of force. Thus:

$$\text{work} = 19.62 \text{ N} \times 1 \text{ m}$$
$$= 19.62 \text{ Nm}$$

The SI unit for work is the Joule (J). *Table 1* indicates that 1 Nm = 1 J, therefore the work performed in the example above was 19.62 J.

Traditionally in the exercise physiology laboratory **force** was quantified in kiloponds (kp), where 1 kp represents the force exerted by 1 kg mass in normal gravity (i.e. 1 kg = 1 kp).

Consequently, moving the 2 kg mass over 1 meter would require the following work to be performed:

$$\text{work} = 2 \text{ kp} \times 1 \text{ m}$$
$$= 2 \text{ kpm (kilopond metres)}$$

From *Table 1* we can see that 1 kpm, or 1 kgm = 9.81 Joules, so 2 kpm is equal to 19.62 J of work done. These examples clearly show how the exercise physiologist can work in many units but must report the final number in the correct SI units.

Table 1. Common conversion factors required in the exercise physiology laboratory

To convert	Into	Conversion factor
mph	m.min^{-1}	× 26.8
min/mile	mph	60 ÷ min/mile
weight in lb	mass in kg	÷ 2.2
power in Watts	workload in kg m min^{-1}	× 6.12
mass in kg	force in Newtons	× 9.81
kgm	Joules	× 9.81
VO$_2$ in L min^{-1}	kcal	× 5
kcal	Joules	× 4186
kcal	Kilojoules	× 4.186
VO$_2$ in ml.kg.min^{-1}	METS	÷ 3.5

Power

Power is defined as the rate at which work is performed, and thus can also be termed **work rate**. This means that time is important in the calculation, and power can be defined as:

$$\text{power} = \frac{\text{work}}{\text{time}}$$

Thus the performance of 150 kpm (or kgm) of work in 1 min would produce a power output of:

$$\text{power} = \frac{150}{1}$$
$$= 150 \text{ kpm min}^{-1} \text{ or kgm min}^{-1}$$

The SI unit for power is Watts (W). *Table 1* provides the conversion factor and thus a power output of 150 kgm min^{-1} ÷ 6.12 = 24.51 W min^{-1} in the above example.

A3 WORK AND POWER PERFORMED ON THE CYCLE ERGOMETER AND TREADMILL

Key Notes

Cycle ergometer	Work and power performed on the cycle ergometer can be calculated by knowing the resistance applied to the flywheel and the distance the flywheel moves per revolution whilst pedalling.
Treadmill	Work performed on the treadmill can be calculated for walking or running up a gradient. The vertical displacement of the body whilst walking or running on a horizontal treadmill is very difficult to measure. Consequently, calculating work for horizontal treadmill movement is complicated.

Cycle ergometer

Work and **power** can be calculated for a participant cycling on the cycle ergometer by using the number of flywheel revolutions per minute, the distance the flywheel travels per revolution and the resistance (or force) applied to the flywheel usually from weights hung from the flywheel belt. The most commonly used cycle ergometer is the Monark, which has a known flywheel movement of 6 m per revolution. The Tunturi moves 3 m per revolution.

Thus, **work** and **power** for a participant cycling on a Monark ergometer at 50 rpm for 5 min at a load of 2 kg would be:

$$\text{work} = \text{force} \times \text{distance}$$
$$\text{force} = 2 \text{ kg}$$
$$\text{distance} = 50 \text{ rpm} \times 6 \text{ m (the circumference of 1 revolution)}$$

So:

$$\text{work} = 2 \text{ kg} \times 50 \text{ rpm} \times 6 \text{ m} \times 5 \text{ min}$$
$$\text{work} = 3000 \text{ kg m}^{-1} \text{ or } 29\,430 \text{ Nm or } 29\,430 \text{ J}$$
(see conversion factors in *Table 1*).
$$\text{power} = 3000 \text{ kg m}^{-1} \div 5 \text{ min} = 600 \text{ kg m}^{-1} \text{ min}^{-1}$$
or 98 W

Treadmill

Work can be calculated for a participant moving on a treadmill by knowing the participant's body weight, and the percentage grade and speed at which the participant is moving. These values allow for the calculation of vertical displacement of the body during movement.

$$\text{vertical displacement} = \% \text{ grade} \times \text{distance travelled}$$

$$\text{distance travelled} = \text{speed of treadmill (m min}^{-1}) \times \text{total min of exercise}$$

Thus the work performed by a participant of 70 kg body mass, running at a treadmill speed of 150 m min^{-1} up a 5% gradient for 10 min would be:

$$\text{vertical displacement} = 0.05 \times 1500 \text{ m}$$
$$= 75 \text{ m}$$

$$\text{work} = 70 \text{ kg} \times 75 \text{ m}$$
$$= 5250 \text{ kg m}^{-1} \text{ or } 15\,102.5 \text{ J}$$

Calculation of work performed during horizontal movement is complicated because the vertical displacement of the body is not easy to measure.

A4 ESTIMATION AND MEASUREMENT OF ENERGY EXPENDITURE

Key Notes

Exercise intensity	The intensity of exercise being undertaken is reported as a measure of work. Intensity is often reported as a percentage of maximal capacity and this capacity may be reported as maximal heart rate, maximal voluntary contraction, or maximal oxygen utilization. Other definitions may use terms such as 'light,' moderate' and 'heavy,' or 'submaximal,' 'maximal' and 'supramaximal,' or above and below the lactate threshold.
Energy expenditure	The measurement or estimation of energy expenditure allows the exercise physiologist to assess the physiological cost of producing physical work.
Oxygen cost	The oxygen cost of performing work (VO_2) is dependent upon the work rate. The oxygen cost can be measured through indirect calorimetry in the exercise physiology laboratory and is reported in absolute ($L\ min^{-1}$) or relative terms ($ml\ kg^{-1}\ min^{-1}$).
Caloric cost	The caloric cost of performing work can be estimated from the measurement of the oxygen cost. One liter of oxygen utilized during exercise is equal to approximately 20 kJ (5 kcal) in energy expenditure. The correct SI units are Joules.
Metabolic equivalent	The metabolic equivalent (MET) provides a generic method of expressing energy expenditure. One MET is equal to resting oxygen consumption (approximately $3.5\ ml\ kg^{-1}\ min^{-1}$). So a subject utilizing $35\ ml\ kg^{-1}\ min^{-1}$ is said to be exercising at 10 METs.

Related topics	Energy for various exercise intensities (B5)	Cardiovascular responses to exercise (E3)
	Pulmonary responses to exercise (D3)	Energy balance (J1)
		Screening and exercise testing (L1)

Exercise intensity   The ability to perform physical work is dependent upon the ability of the muscle to transform chemical energy into mechanical energy. In order to do this the muscle is also dependent upon the ability of the circulatory system to deliver oxygen to the tissue, and the ability of the tissue to extract and utilize the delivered oxygen. Consequently, a person could be working at a **work rate** that elicits the maximal capacity of, or 50% of the maximal capacity of, the circulatory system or oxygen utilization. Either way exercise physiologists refer to this work rate, or this percentage of maximal capacity, as **exercise intensity**.

The American College of Sports Medicine classifies exercise intensity as either a percentage of maximal oxygen uptake or maximal heart rate, or in metabolic equivalents (METs). So, for instance, a subject can be exercising at 50% of their maximal oxygen uptake (the maximal amount of oxygen the body can utilize during exercise at sea level) or maximal heart rate (see Section L). METs are described below. Other exercise domains are described for work undertaken above or below the lactate threshold (the point at which lactate begins to accumulate in the blood: see Section B5). Work rates below the lactate threshold are referred to as **moderate exercise**, whilst those above the lactate threshold but below maximal oxygen uptake are referred to as being **heavy exercise**. Within this book work rates above that associated with maximal oxygen uptake will be referred to as '**supramaximal'** (e.g. sprints).

Energy expenditure

Measurement of **energy expenditure** allows the exercise physiologist to calculate the metabolic cost of exercise (see also Section J1). Energy expended by the body is reflected as heat production, and this heat production is quantified in **calories**. The kilocalorie (1000 calories) is the amount of heat required to raise the temperature of 1 kg of water by 1°C. Consequently, assessing energy expended during exercise in **kilocalories**, or the SI units of **Joules**, provides a measure of the physiological cost of producing physical work. Heat produced during metabolism can be measured in a whole body chamber by **direct calorimetry**.

Oxygen cost

The **oxygen cost** of performing work ($\dot{V}O_2$) is primarily dependent upon the work rate. As work rate increases $\dot{V}O_2$ increases linearly for moderate work rates (those below the lactate threshold). The **oxygen cost** is reflective of the metabolic cost of the exercise (see Section B) and is usually measured in the laboratory using **indirect calorimetry**. Expired air is collected in **Douglas bags**, or sampled directly with an on-line gas analyzer, and the fractions of expired oxygen (F_EO_2) and carbon dioxide (F_ECO_2) in the expired air measured. The volume of air expired per minute is usually measured via a dry gas meter, and is then recorded as the **minute ventilation**, or volume expired (VE). Dry gas meters usually possess internal thermometers in order that the temperature of the expired air can be recorded.

In the laboratory an experiment might involve a subject cycling at 100 W for 10 minutes whilst expired air is collected during the final minute of exercise. When the air in the Douglas bag has been sampled for O_2 and CO_2, and the quantity of air, and temperature of air in the bag have been measured, all the student now requires in order to complete the calculation of oxygen uptake is the room temperature and atmospheric temperature, pressure and relative humidity.

As any volume of air is influenced by atmospheric temperature, pressure and water vapor, it is important that volumes are reported as a standard against which all laboratories can compare results. So, oxygen uptake is calculated using the volumes that were recorded at atmospheric (A) temperature (T) and pressure (P). These volumes would also be saturated (S) with water vapor. Thus they are said to be measured at ATPS. In order to calculate oxygen uptake these volumes need to be corrected to a standard (S) temperature (T: 273 K or 0°C), standard pressure (P: 760 mmHg), and the volume that they would be in dry air (D). Thus the corrected volume is said to be recorded as STPD.

The following equation is used to correct the VE (ATPS) to VE (STPD):

$$\text{VE (STPD)} = \text{VE (ATPS)} \times \left(\frac{273}{273 + T}\right) \times \left(\frac{PB - PH_2O}{760}\right)$$

where T = temperature of gas in measuring device (°C)
PB = atmospheric pressure (mmHg)
PH_2O = partial pressure of water (from table in most physiology books)

Following this correction $\dot{V}O_2$ is calculated using the VE (STPD) as:

$$\dot{V}O_2 = \text{VE}[(\%N_E \times 0.265) - FEO_2 \%]$$

where $\%N_E$ is the % nitrogen expired. This is calculated as:

$$\%N_E = [100 - (\%O_{2E} + \%CO_{2E})]$$

Absolute oxygen cost is expressed in liters or milliliters of oxygen consumed per minute (L min^{-1} or ml min^{-1}), whilst **relative oxygen cost** is expressed in milliliters of oxygen consumed per kilogram of body mass per minute (ml kg^{-1} min^{-1}). During weightbearing activity such as running, the oxygen cost of the activity is expressed in relative terms. If the weight is supported, as in cycle ergometry, it is usual to report oxygen cost in absolute terms.

The **total oxygen cost** of movement incorporates the oxygen cost of rest, the oxygen cost of moving the legs in cycle ergometry, or the whole body in tread-mill exercise, plus the oxygen cost of performing work. The **true oxygen cost** of performing work is thus the oxygen cost above that of rest and leg or body movement. On the cycle ergometer the oxygen cost of moving the legs is meas-ured during cycling against a zero load (0 W). The oxygen cost of performing the additional work is then measured as the difference between the oxygen uptake measured at a given work rate and that recorded at 0 W.

The maximal rate that the body can consume oxygen during physical activity at sea level is termed **maximal oxygen consumption** or **uptake**. This value can be measured from a maximal oxygen uptake exercise test, or predicted from a submaximal test (see Section L). Maximal tests involve incremental activity, on any ergometer, whereby the subject exercises at increasing work rates until voli-tional fatigue. Oxygen consumption and heart rate are measured and recorded during the final minute of each increment. Traditionally maximal oxygen uptake ($\dot{V}O_2$max) is verified by

- no further increase in $\dot{V}O_2$ despite an increase in work rate
- a maximal heart rate within 10 beats min^{-1} of the predicted maximum (220 – age)
- a respiratory exchange ratio (see Section B) of 1.1 or above
- blood lactate concentration exceeds 8–9 mM (see Section B).

Maximal oxygen uptake has been thought of as the major variable indicating the functional capacity of the cardiorespiratory system, and depending upon the initial level of fitness can be increased by physical training. Average values for men and women are approximately 30–55 ml kg min^{-1} and 25–40 ml kg min^{-1} respectively. Endurance-trained athletes may have a $\dot{V}O_2$max of greater than 80 ml kg min^{-1}.

Caloric cost

The **caloric cost** of exercise provides an estimation of the metabolic energy utilized in producing skeletal work. The cost can be estimated from the oxygen cost of work in that one liter of oxygen consumed represents approximately

5 kcal (20 kJ) of energy expenditure. In reality this figure is dependent upon substrate utilization (see Section B). However, the approximation of 5 kcal (20 kJ) allows the exercise physiologist to estimate the caloric cost of an exercise session and compare it to the measured caloric intake in diet. Weight-loss programs are designed and assessed using this protocol (see Section J).

Metabolic equivalent

The **metabolic equivalent** (**MET**) provides a generic unit of energy expenditure. One MET represents resting energy expenditure and is equivalent to 3.5 ml kg^{-1} min^{-1} of oxygen consumption. Consequently, if a person is consuming 35 ml kg^{-1} min^{-1} of oxygen during an exercise task he/she can be reported to be working at 10 METs.

B1 ENERGY SOURCES AND EXERCISE

Key Notes

The energy continuum	The energy continuum is a schematic which highlights the major energy systems involved when exercise progresses from short, high-intensity exercise to more prolonged, lower-intensity exercise. The energy sources for the high-intensity exercise are mainly derived from anaerobic sources whereas low-intensity exercise derives its energy mainly from aerobic processes.
ATP	Adenosine triphosphate (ATP) is the energy currency used by muscle cells to allow them to produce force and is broken down to form adenosine diphosphate (ADP).
Creatine phosphate	Creatine phosphate or phosphocreatine (PCr) is a high-energy phosphate molecule found in cells which is an immediate source of re-forming ATP from ADP.
Muscle glycogen	Muscle glycogen is the storage form of carbohydrate, and is made up of glucose molecules. Glycogen can be broken down rapidly to produce ATP for intense exercise or more slowly for prolonged exercise.
Glycogenolysis and glycolysis	Glycogenolysis is the process of removing glucose subunits from a glycogen molecule. The enzyme, glycogen phosphorylase, breaks off glucose molecules to form glucose-1-phosphate. Glycolysis is the process of converting glucose to pyruvic acid in the cytoplasm of cells, with a net production of ATP. The process does not require oxygen, and whereas during steady-state exercise most of the pyruvic acid is processed through aerobic breakdown, during high-intensity exercise the resultant formation of lactic acid occurs.
Blood glucose	Blood glucose is a carbohydrate source of energy for cells. The glucose in blood arises from the liver.
Aerobic energy systems	Aerobic energy is produced in the mitochondria of cells. During prolonged exercise, the two major sources are carbohydrates and lipids.
Related topics	Fiber types (C4) Macronutrients (G2)

The energy continuum

Examination of the major energy contribution from varying sources as the duration of exercise progresses can be seen in *Fig. 1*. At the start of an activity the initial source of energy is from the adenosine triphosphate (ATP) stores at the muscle crossbridges. The ATP, when used, is broken down to adenosine diphosphate (ADP). Creatine phosphate (PCr) rapidly replaces the ATP and so, in an indirect sense, becomes the next major source of energy. The anaerobic breakdown of muscle glycogen through glycolysis to form ATP and lactic acid

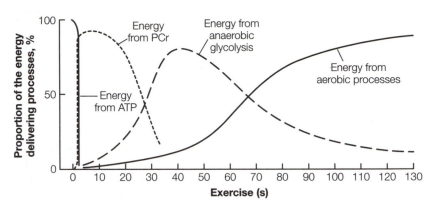

Fig. 1.　Schematic of the energy continuum. (It is important to note that each of these mechanisms of energy production occur simultaneously.)

predominates after PCr, whilst aerobic processes predominate from approximately 60 s and beyond. The **energy continuum** depicts the changes in major energy sources with time when exercise is maximal for each of the phases. For example at 20 s of exercise at maximal intensity, approximately 40% of the energy derives from PCr, 50% from glycolysis, and 10% from aerobic processes, whereas at 40 s the contribution is 5%, 80%, and 15% for PCr, glycolysis and aerobic processes, respectively. Understanding of the energy continuum enables coaches and athletes to appreciate the major energy sources being used when exercising maximally in short or repetitive sprints during training or in games.

ATP　　　　　**Adenosine triphosphate (ATP)** is a ubiquitous high-energy phosphate which consists of a nucleoside, adenosine, to which is attached three phosphate molecules using high energy-yielding bonds (*Fig. 2*). When a phosphate is removed from ATP, the energy produced provides the currency to enable muscles to move, molecules to be synthesized or to be transported against a concentration gradient or to be excreted. Indeed, any energy requiring processes invariably uses ATP as the prime source of energy. ATP is hydrolyzed by enzymes called **ATPases** which result in the formation of **adenosine diphosphate (ADP)** and **inorganic phosphate (P_i)**. The equation can be represented as follows:

$$\text{adenosine} - P - P - P \longleftrightarrow \text{adenosine} - P - P + P_i + \text{energy (37 kJ)}$$

$$\uparrow$$

$$\textbf{ATPase}$$

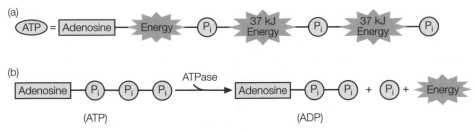

Fig. 2.　Schematic of (a) the structure and, (b) hydrolysis of ATP.

The amount of ATP in muscle is rather small, with the concentration being approximately 20–30 mM kg^{-1} of dry muscle. This amount of ATP in muscle has been estimated to be sufficient to fuel around 3–5 s of maximal effort if ATP was the sole energy source. Clearly this is not the case, because maximal efforts last longer, and so the restoration of ATP must occur.

Creatine phosphate

The other immediate source of energy for high intensity exercise is that of **creatine phosphate** or **phosphocreatine (PCr)**. PCr is also a high-energy phosphate in which a single phosphate molecule is attached to a molecule of creatine. Hydrolysis of PCr is brought about by the enzyme **creatine kinase (CK)** as follows:

$$\text{creatine} - \text{P} \longleftrightarrow \text{creatine} + \text{P}_i + \text{energy (43 kJ)}$$
$$\uparrow$$
$$\textbf{creatine kinase}$$

The reaction above is linked to the re-forming (rephosphorylation) of ATP from ADP:

$$\text{PCr} + \text{ADP} \longleftrightarrow \text{Cr} + \text{ATP}$$
$$\uparrow$$
$$\textbf{CK}$$

The enzyme CK exists in a number of **isoenzymes**, which have the same formula but possess different rates for the above reaction. The two best examples from a muscle and exercise context are CK_{mm} (the form of creatine kinase found at the muscle crossbridge) and CK_{mito} (the mitochondrial form of creatine kinase). Whereas CK_{mm} favors the above reaction from left to right, the isoform CK_{mito} favors the reaction from right to left. The hydrolysis of PCr occurs during intense bouts of exercise at the crossbridges whilst resynthesis of PCr from creatine and ATP occurs during recovery phases at the mitochondrial membrane (*Fig. 3*).

The total creatine pool within muscle represents around 120 mM kg^{-1} dry muscle of which PCr is 80% under normal resting conditions (i.e. 96 mM kg^{-1} dm) and actual Cr concentration is about 20% (i.e. 24 mM kg^{-1} dm). Following

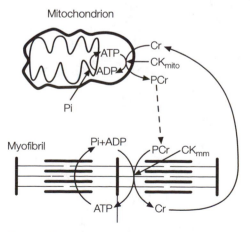

Fig. 3. Schematic of the use and resynthesis of creatine phosphate.

strenuous exercise the PCr and Cr contents will change significantly. This amount of PCr is sufficient to fuel maximal exercise solely for approximately 6–8 s. Section G explores the potential of creatine supplementation as an ergogenic aid.

Muscle glycogen

Muscle glycogen is an essential store of carbohydrate fuel for both high-intensity exercise and also prolonged activity. Glycogen is a **polysaccharide** made up of a large number of **monosaccharide** glucose units joined together by two types of bonds i.e. 1,4-α bonds and 1,6-α bonds. The former produce straight chains of glucose molecules but after every ten glucose units or so, the chain is branched by the 1,6 bond. *Fig. 4* illustrates the branching nature of glycogen. (For more detail see *Instant Notes in Biochemistry*, Chapters J1 and J2.)

Normal muscle glycogen stores have a concentration of 350 mM kg^{-1} dry muscle, although this can be increased significantly by a high carbohydrate diet and reduced significantly by repeated bouts of sprinting or a single prolonged bout of exercise or when on a low carbohydrate diet for around 2–4 days.

Glycogenolysis and glycolysis

When muscle glycogen breaks down to produce energy, the reactions involve the removal of glucose molecules under the influence of the enzyme **glycogen phosphorylase**. The process of breaking off glucose molecules from glycogen is known as **glycogenolysis**. The initial glucose product is glucose-1-phosphate, which is then converted to glucose-6-phosphate. Once G-6-P has been produced, there is a common pathway through **glycolysis** from the glucose removed from glycogen and glucose entering the cell from blood. Glycolysis is a series of processes which takes place in the cytoplasm of cells resulting in the formation of two pyruvic acid molecules and ATP. If the activity is intense,

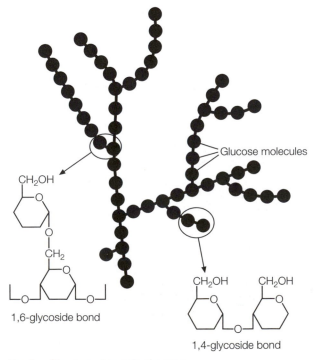

Fig. 4.   Structure of muscle glycogen.

the pyruvic acid leads to the formation of **lactic acid**, although during steady-state exercise the majority of the pyruvic acid formed is broken down via the **aerobic** pathway to produce carbon dioxide and water. *Fig. 5* shows a schematic of the overall processes of glycogenolysis and glycolysis.

Blood glucose

Glucose delivered to the muscle by blood may also act as a useful energy source during exercise. The normal **blood glucose** concentration is around 5 mM, but can be elevated to values in excess of 7–8 mM (**hyperglycemia**) following a high carbohydrate meal or reduced to values lower than 4 mM (**hypoglycemia**) if exercising for prolonged periods of time without carbohydrates being ingested. The glucose in blood is produced by the liver either from glycogenolysis of its own glycogen stores or from **gluconeogenesis**, where the glucose is produced from precursors such as lactic acid, alanine, pyruvic acid, or glycerol.

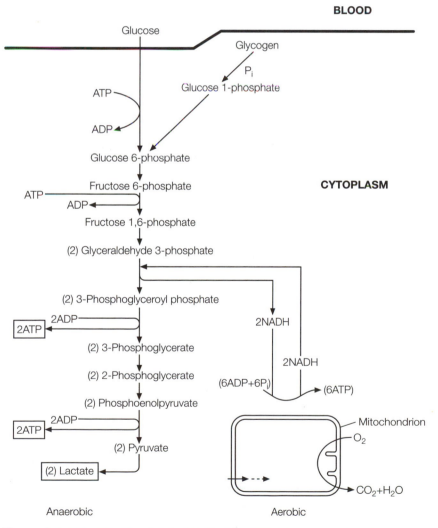

Fig. 5. Schematic showing overall processes of glycogenolysis and glycolysis. (2) = 2 molecules

**Aerobic energy
systems**

Aerobic energy processes take place in the mitochondria of cells. The essential
requirement is that oxygen must be present to complete the process. During
steady-state exercise the pyruvic acid produced is converted to **acetyl-CoA** in
the mitochondria and then undergoes aerobic oxidation via the **TCA cycle**. Fats
are another major source of energy during prolonged exercise and can only be
used to produce energy using aerobic processes. Essentially fats are converted,
through **β-oxidation**, to produce acetyl-CoA and then enter the TCA cycle.
Hence acetyl-CoA is an important crossroad in carbohydrate and fat metabolism. *Fig. 6* presents an overview of aerobic processes.

For further details on the biochemistry of these processes you would be
advised to consult *Instant Notes in Biochemistry*.

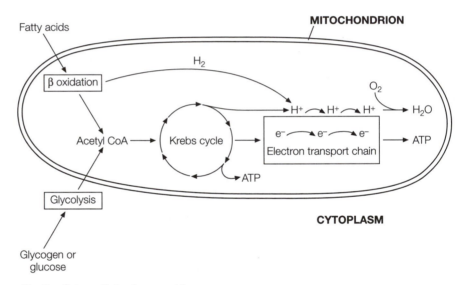

Fig. 6.   Schematic to show aerobic processes.

B2 RATES OF ENERGY PRODUCTION

Key Notes

Rate of ATP production and power	The maximum power generated during exercise is dependent on the maximal rates of ATP utilized by muscle. The energy sources include ATP, PCr, glycolysis, and the aerobic oxidation of carbohydrates and lipids.
ATPase, creatine kinase and glycolysis	The maximal rate of ATP use is dependent on the maximal rate of ATPase activity at the cross-bridges. The maximal rate of ATP generation if employing PCr is dependent on the maximal rate of creatine kinase activity. The maximal rate of power generated through glycolysis is dependent on the maximal rate of the rate-limiting enzyme. In the case of glycolysis, the rate-limiting enzyme is PFK.
TCA cycle	The maximal rate of energy production using the aerobic breakdown of carbohydrates is dependent on the slowest rate of ATP production from glycolysis or the TCA cycle. It appears that ATP production through the TCA cycle is approximately half that generated via glycolysis. The maximal rate of ATP generated through aerobic processes when using lipids is half that of carbohydrates. The problem of 'hitting the wall' can be explained by the lack of carbohydrate and the fact that muscle then has to rely exclusively on lipids.
Related topics	Fiber types (C4) Nutrition and ergogenic aids for
	Integrated control of exercise (F3) sports performance (Section G)

Rate of ATP production and power

The maximal rate of ATP production is related to the power developed during a bout of exercise and is dependent on the maximal rates of ATP utilized by the muscle. These energy sources include those contained within the energy continuum, i.e. ATP, PCr, breakdown of muscle glycogen rapidly resulting in lactic acid formation, and the aerobic oxidation of carbohydrates and fats. In a test of power such as the **Wingate test**, the power produced is due to the rapid energy production from four sources of energy which will be highlighted below. Examination of a typical power curve produced when undertaking the Wingate test (*Fig. 1*) shows peak power within the first few seconds followed by a gradual decline in power production. The major sources of energy include ATP and PCr during the first 6–10 seconds followed by **anaerobic glycogenolysis** and then aerobic processes (*Fig. 1*).

Each of the energy-producing processes involved in the generation of power for a Wingate test has maximal rates of ATP production. These figures can be seen in *Table 1*.

Experiments which have used **muscle biopsy** before and after maximal-intensity exercise of varying durations have resulted in the data obtained for *Table 1*. In order to achieve this, a subject would have a muscle biopsy undertaken at rest and this would be followed by a further biopsy taken immediately after an all-out bout of exercise lasting for between 6 and 60 s. The depletion of the

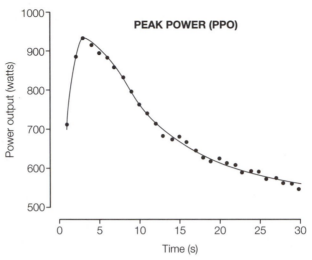

Fig. 1.   Typical power profile from a Wingate test.

Table 1. Rate of ATP production from various energy processes

Process	Rate of ATP production (mM kg^{-1} dry muscle/second)
ATP $\leftrightarrow$ ADP + P$_i$	12
PCr $\leftrightarrow$ Cr + P$_i$	8–10
Glycogen $\rightarrow$ lactic acid	4
Glycogen/glucose $\rightarrow$ CO$_2$ + H$_2$O	2
Lipid $\rightarrow$ CO$_2$ + H$_2$O	1

energy sources over the time periods can be employed to reflect the rate of use of that source, e.g. after 6 s of sprint cycling the depletion of ATP has been found to be 9% and that for PCr to be 35%. Since the exercise lasts for 6 s the rate of use may be calculated. Likewise, this process can be repeated for bouts lasting 10, 20, 30 and 60 s.

ATPase, creatine kinase and glycolysis

The maximal rate of ATP utilization during intense exercise is approximately 12 mM kg dry muscle^{-1} s^{-1}, and reflects the rate of ATP hydrolysis by ATPases at the cross-bridges. The maximal rate of ATP production from PCr is lower than that for ATP hydrolysis, and is limited by the maximal rate of PCr degradation by CK. Interestingly, the maximal rate of ATP use from anaerobic glycolysis is approximately half that for PCr degradation. This is because the rate-limiting enzyme for glycolysis is **PFK (phosphofructokinase)**, which has a lower activity than CK.

TCA cycle

The two major aerobic energy processes shown in *Table 1* (i.e. muscle glycogen and lipids) have maximal rates of ATP production much lower than the anaerobic sources. Although these rates of production are relatively low, the total amount of energy from these aerobic stores is larger. From *Table 1* it is evident that carbohydrates can produce ATP at twice the rate that fats can be maximally oxidized. Since there is a common oxidative process from acetyl-CoA onwards for both carbohydrates and fats (i.e. **TCA cycle**), the slower rates of ATP production from fats must be as a result of slower rates of entry into the muscle cell or uptake across the mitochondria, or that β-oxidation is slower than TCA cycle activity.

B3 ENERGY STORES

Key Notes

Carbohydrates	Carbohydrates are stored mainly in muscle and liver as glycogen. The total amount is approximately 100–150 g in the liver and 200–300 g in muscle, although diet and exercise can alter these amounts substantially.
Lipids	Lipids are mainly stored in adipose tissue as triglycerides, a combination of three fatty acids with a glycerol molecule. However, muscle triglyceride stores are also present. The total amount of lipids is significant, with values dependent on the % body fat of an individual. The % body fat can vary between 5% in elite male distance cyclists to values greater than 30% in some obese individuals.
Protein	Muscle is the largest source of protein that can be used as a fuel during exercise. The total amount available is therefore dependent on the muscle mass.

Related topics   The endocrine system (F2)   Energy balance (J1)
                     Nutrition and ergogenic aids for
                         sports performance (Section G)

Carbohydrates   Carbohydrates are an important energy source for both intense and prolonged exercise, and are stored as **glycogen** in both the muscle where it is needed as an energy source and in liver where the glucose is pushed out into the blood. In comparison with lipid stores, carbohydrates are limited with a total amount of approximately 500 g in the body. The muscle contains around 300 g of glycogen, although this store can increase to 500 or 600 g when carbohydrate loaded, whilst the liver glycogen stores are about 150 g in total. The liver glycogen stores are also affected by diet and exercise in so far as these stores are enhanced by high carbohydrate feeding and depleted by either prolonged exercise or fasting. Indeed, after an overnight fast the liver glycogen stores can be severely depleted.

The amount of energy contained within the carbohydrate stores can be estimated by multiplying the total amount in grams by 3.75 or 16, which are the energy content of a carbohydrate in kcal or kJ respectively. Therefore a 300 g muscle glycogen content contains 1125 kcal or 4800 kJ of energy.

Glycogen is produced from glucose carried to the muscle or liver in blood and is regulated by the hormone, **insulin**. Normally this process arises after ingesting a meal containing carbohydrates. Increased levels of circulating blood insulin result in enhanced uptake of glucose by muscle and the subsequent packaging of the glucose to existing glycogen to make a larger glycogen molecule. The key enzyme responsible for this process of **glycogenesis** is **glycogen synthase**. *Fig. 1* illustrates how glycogenesis occurs. The enzyme glycogen synthase works in opposition to the enzyme glycogen phosphorylase, and so

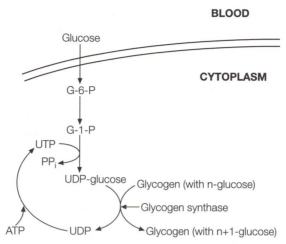

Fig. 1.   Schematic to show the process of glycogenesis.

any factors which are likely to promote the activity of glycogen phosphorylase will inhibit glycogen synthase and vice versa.

Lipids

Lipid stores are, in contrast to carbohydrates, a major store of energy, since each fat molecule contains 9 kcal (38 kJ) of energy. This is more than twice the **energy density** found in carbohydrates and so makes lipids a useful, compact storage source. Lipids are stored as **triglycerides**, which are essentially a glycerol molecule that is attached to three fatty acids (*Fig. 2*). Major stores of triglycerides can be found in adipose tissue, although they may also be found inside muscle cells. Although triglycerides are the storage form of lipids, fatty acids are the useable form from a metabolic perspective.

Triglycerides are produced when there is an excess amount of fat intake or when there is an excessive amount of carbohydrate ingested. The **fatty acids** from the digested food become attached to a glycerol molecule to produce triglycerides. In the case of excessive amounts of carbohydrate ingested, the glucose is used to produce both **glycerol** and fatty acids, which together make up a triglyceride molecule.

To estimate the total amount of energy in fat stores it is first necessary to determine how much fat there is. This can be done by estimating the % body fat using either skinfold measures or **DEXA** scan or by **underwater weighing**. The average % body fat for young males varies between 10 and 20% and for females between 20 and 30%. If we take a male weighing 70 kg and with an estimated body fat of 15%, then the estimated total fat content is 10.5 kg (i.e. 15% of

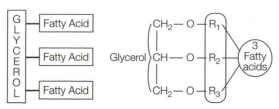

Fig. 2.   Triglyceride molecule.

70 kg). To find out the energy content it is necessary to multiply the fat mass in grams by 9 or 38 to get values in kcal or kJ. Such a calculation shows that the energy content is 94 500 kcal or 399 000 kJ, a plentiful supply of energy. Even in those athletes with extremely low body fat % such as elite male road cyclists with values around 5% body fat, there is sufficient energy to undertake extremely prolonged exercise.

Protein

The major stores of metabolic protein are muscles. Of course proteins include all enzymes, peptide hormones, molecules such as **hemoglobin** and **myoglobin**, and are an integral part of all membranes. However, these sources are not broken down to be used as energy. The total mass of potentially usable protein is about 8 kg, although this depends on the muscle mass of an individual. Protein provides 4 kcal (17 kJ) of energy per gram, and so if there were 8 kg of protein, this would provide 32 000 kcal (136 000 kJ) of energy. The problem with using protein as an energy source is that it means the muscle is cannibalizing itself to provide energy.

B4 CONTROL OF ENERGY SOURCES

Key Notes

Metabolic regulation
Metabolism is regulated through control of key enzymes involved in energy-producing or energy-storage processes. The control is mediated by hormones which, via cAMP, activate inactive enzymes. However, enzyme activity can also be affected by allosteric effectors.

Hormones and cyclic AMP
Hormones are chemical substances originating from glandular cells which are then transported through blood to a target cell to influence physiological activity. Examples of key hormones which influence the metabolic activity of a cell include adrenaline (epinephrine), noradrenaline (norepinephrine), insulin, and glucagon.
  Cyclic AMP (cAMP) is a ubiquitous intracellular effector molecule produced from ATP by the enzyme adenylate kinase. Cyclic AMP activates a protein kinase within cells, which in turn activates inactive enzymes. In some instances (e.g. glycogen synthase) cAMP production inhibits the activity of an enzyme. Production of cAMP occurs as a consequence of hormonal activation.

Allosteric effectors
Some molecules found in cells are able to promote or inhibit the activity of regulatory enzymes. Such molecules, known as allosteric effectors, do not bind to the active site of an enzyme and as such are distinct from competitive inhibitors.

Related topics
The endocrine system (F2)   Integrated control of exercise (F3)

Metabolic regulation
To ensure that ATP is provided rapidly or slowly, there needs to be regulation of the metabolic processes. Such regulation is normally controlled during steady-state exercise by circulating hormones secreted by **endocrine** glands such as the adrenal glands or the pancreas. The hormones secreted by these glands include **adrenaline (epinephrine), noradrenaline (norepinephrine), glucagon, and insulin**, and they mediate their effects by activating inactive enzymes present within the cell. However if the exercise is intense and rapid, there may be no opportunity for hormones to regulate such activity as energy is required within seconds. Under these circumstances the switching on of energy processes must take place due to activation within the cell.

Hormones and cyclic AMP
There are essentially three classes of hormones, these being **amine hormones**, **peptide hormones** and **steroid hormones**. The amine hormones include the catecholamines adrenaline (epinephrine) and noradrenaline (norepinephrine), the peptide hormones include glucagon and insulin, whereas the steroid hormones include the sex hormones estrogen and testosterone. Peptide and amine hormones affect their target cells by attaching to a receptor on the plasma membrane and switching on the production of **cyclic AMP (cAMP)**.

Cyclic AMP is ubiquitous and in the first instance usually activates an inactive **protein kinase**, which in turn activates various inactive enzymes in the target cell. Once the inactive enzymes are activated, they undertake their metabolic role. *Fig. 1* illustrates how adrenaline (epinephrine) activates the enzyme glycogen phosphorylase to break down glycogen, whilst *Fig. 2* shows how breakdown of **triglyceride** stores in adipose tissue is also regulated by adrenaline (epinephrine) via production of cAMP. The breakdown of triglycerides leads to the formation of fatty acids and glycerol, and is known as **lipolysis**.

The key hormones concerned with regulating energy metabolism are the amine and peptide hormones, although the steroid hormones cortisol and growth hormone also affect metabolic processes. These hormones have as their target tissues either muscle, adipose tissue or liver, and regulate the processes glycogenolysis, glycolysis, lipolysis, gluconeogenesis, and glycogenesis, as well as protein degradation and synthesis. *Table 1* shows which processes are affected by the hormone in these tissues.

Table 1.   Hormonal influence on energy metabolic processes in varying tissues

Hormone	Target tissue and process regulated		
	Muscle	Adipose	Liver
Adrenaline (epinephrine)	Glycogenolysis ↑ Glycolysis ↑	Lipolysis ↑	Glycogenolysis ↑
Noradrenaline (norepinephrine)		Lipolysis ↑	
Insulin	Glycogenesis ↑ Protein synthesis ↑	Lipolysis ↓ Lipogenesis ↑	
Glucagon	Protein degradation ↑		Glycogenolysis ↑ Gluconeogenesis ↑
Cortisol		Lipolysis ↑	
Growth hormone	Protein synthesis ↑	Lipolysis ↑	

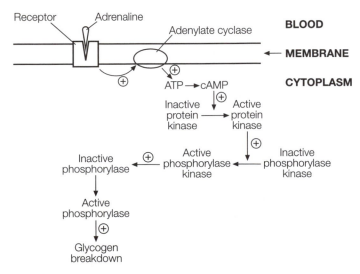

Fig. 1.   Schematic to show regulation of glycogen phosphorylase activity.

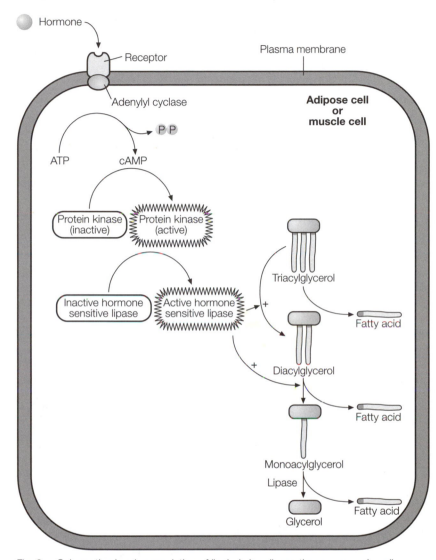

Fig. 2.   Schematic showing regulation of lipolysis in adipose tissue or muscle cells.

Allosteric effectors

The metabolic processes involved in energy production and storage are regulated by hormones. The question arises as to whether hormones influence metabolic processes fast enough to generate ATP during short, intense activity. Under such circumstances it is unlikely that hormones such as adrenaline (epinephrine) can switch on glycogenolysis quickly enough, and so there is a need to switch on glycogen phosphorylase more rapidly. This is achieved by the release of **calcium (Ca^{2+})** ions from the **sarcoplasmic reticulum** during muscle contraction, and the Ca^{2+} directly activates glycogen phosphorylase. Such an activation (or inhibition) of an enzyme by a cell product is an example of **allosteric regulation**. Allosteric regulators include ATP, ADP, ammonia, citric acid, and products of the process being catalyzed by the enzyme itself.

Two examples of allosteric effectors can be seen in *Fig. 3*, where Ca²⁺ can be seen to affect glycogen phosphorylase and how ATP, ADP, **ammonia, citric acid,** and fructose 1,6 diphosphate affect PFK (the regulatory enzyme for glycolysis).

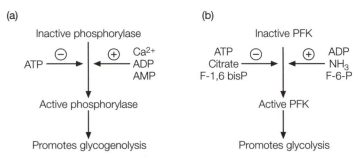

Fig. 3. Examples of allosteric effectors: (a) control of glycogen phosphorylase, (b) control of PFK activity.

B5 ENERGY FOR EXERCISE OF VARYING INTENSITIES

Key Notes

High-intensity exercise	Carbohydrates are the only macronutrients that can be used as an energy source during high-intensity exercise. Muscle glycogen is broken down in the cytoplasm during intense exercise to produce energy and lactic acid.
Prolonged steady-state exercise	Muscle and liver glycogen stores are used to provide energy during prolonged exercise. Depletion of liver glycogen can result in hypoglycemia, whereas depletion of muscle glycogen results in the inability to sustain an exercise intensity above 55% $\dot{V}O_{2max}$.
Intermittent exercise	Carbohydrates are an important source of energy during intermittent bouts of exercise, in particular the more intense intervals.

Related topics	Fiber types (C4)	Nutrition and ergogenic aids for
	The endocrine system (F2)	sports performance (Section G)

High-intensity exercise

High-intensity exercise is exercise that is non-steady-state exercise, and as such is likely to be maintained for slightly longer than 5 minutes before fatigue ensues. This means that such exercise can be maximal and last for a few seconds or slightly more prolonged. The major sources of energy for high-intensity exercise are ATP, PCr, and muscle glycogen. All these energy sources reside in the cytoplasm and can be used rapidly from processes that take place in the cytoplasm. Fats are unable to be used for such intense exercise because they are only broken down in the mitochondria.

Studies employing muscle biopsy sampling immediately following high-intensity exercise have demonstrated that after 30 s of sprinting 80% of the energy is derived from anaerobic sources. In more detail, these studies have revealed that intense exercise of 6 s durations have resulted in ATP, PCr, and muscle glycogen being depleted by 9%, 35%, and 17% respectively, whilst after 30 seconds of sprinting the reductions are 44%, 66%, and 30% respectively.

Fatigue, which may be defined as the inability to sustain the necessary exercise intensity associated with high-intensity exercise is most often considered to be due to lactic acid accumulation and the associated increase in acidity and reduction in **pH** due to **H⁺ ions**. This is because **lactic acid** dissociates into lactate ions and hydrogen ions, and it is the hydrogen ions that cause the decrease in pH:

$$\text{lactic acid (Hla)} \rightarrow \text{La}^- + \text{H}^+$$

The normal resting pH of muscle is around 7, whereas values as low as pH of 6.4 have been reported in muscle following repeated sprint efforts. A decrease in pH to a value of 6.4 has been shown to impair the activity of some key regulatory enzymes such as glycogen phosphorylase, PFK, and even the ATPases. In addition, the H^+ may directly affect Ca^{2+} at the **crossbridges** by competing with them for binding sites.

In spite of much of the early evidence reported to show an impact of lactic acid accumulation on fatigue, there are several more recent studies which demonstrate that lactic acid is not the only fatiguing factor. Increases in inorganic phosphate (P_i) as well as reductions in PCr are important factors relating to fatigue during high-intensity exercise.

Although lactic acid is considered to be a problematical product of high-intensity exercise, the measure of blood lactic acid concentrations at varying exercise intensities has proved to be a most valuable indicator of aerobic capacity. Depending on the fitness of the individual, most prolonged steady-state exercise is performed at an intensity between 55 and 65% $\dot{V}O_{2max}$. However, many endurance-trained athletes can exercise at intensities of around 75 or even 80% $\dot{V}O_{2max}$ for prolonged periods of time and without accumulating lactic acid. The exercise intensity at which such prolonged exercise can be maintained is often associated with the **lactate threshold**. The lactate threshold is an exercise intensity above which there is a likely gradual increase in lactic acid. *Fig. 1* shows that the lactic acid concentrations in blood remain quite steady until a certain intensity is achieved, and after which there is a non-linear increase. This is known as the lactate threshold. From *Fig. 1* it is also clear that endurance training shifts the graph to the right, i.e. there is an increase in the exercise intensity corresponding to the point where lactic acid begins to accumulate.

The lactate threshold occurs for a number of reasons.

1. As exercise intensity increases so does the recruitment of fast glycolytic fibers which favor lactate production.
2. As exercise intensity increases the rate of glycolysis increases above that of the TCA cycle and hence more of the **pyruvic acid** formed is converted to lactic acid.
3. As exercise intensity increases there is a limited delivery of oxygen to muscle and hence a greater chance of **anaerobiosis**.

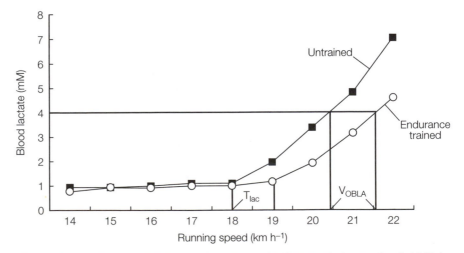

Fig. 1. Blood lactate concentration vs running speed to illustrate the lactate threshold (T_{lac}) and V_{OBLA}.

The reasons why the lactate threshold graph shifts to the right are because endurance training results in:

1. increased lactate removal and reduced lactate production
2. increased **capillarization** of muscle to enable greater oxygen delivery
3. increased **mitochondrial density** and hence more oxidative enzymes to reduce lactate production and enhance lactate removal
4. **slow oxidative** and **intermediate fibers** are particularly influenced in being more oxidative.

Prolonged steady-state exercise

The exercise intensity enabling prolonged activity is dependent on the level of training. The major energy source during such exercise is initially carbohydrate from muscle glycogen and blood glucose, but as the exercise progresses there is a greater dependence on fatty acids either from intramuscular triglyceride stores from **adipose tissue**. *Fig. 2* shows the changes in fat and carbohydrate use during exercise lasting for 60 minutes, and clearly demonstrates the switch from predominant carbohydrate to predominant fat oxidation after around 20 minutes. It takes this long because of the need to stimulate lipolysis and then release the fatty acids under the influence of the hormones adrenaline (epinephrine) and insulin. Increases in adrenaline (epinephrine) and a fall in insulin result in these changes.

A simple way of determining the contribution of carbohydrate and fat to total oxidation is by measuring oxygen uptake and the **respiratory exchange ratio (RER)**. The latter is also referred to as the **respiratory quotient (RQ)**. The distinction between RQ and RER is that the former reflects CO_2 production and O_2 consumption in homogeneous tissue whereas the latter reflects whole body CO_2 production and O_2 consumption.

$$RER = \frac{\text{volume of carbon dioxide produced}}{\text{volume of oxygen consumed}}$$

An RER of 1 reflects exclusive use of carbohydrates whilst an RER of 0.7 reflects only fats being used as an energy source. *Fig. 3* shows how the relationship between RER and exercise intensity affects RER and thereby carbohydrate and fat use.

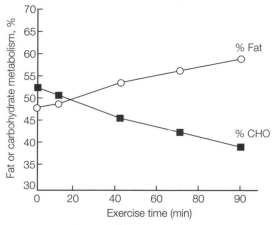

Fig. 2. Contribution of carbohydrate and fat as energy sources during prolonged exercise at 65% $\dot{V}O_{2max}$.

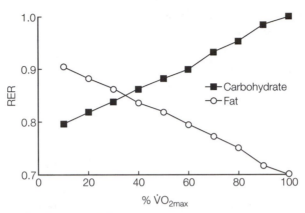

Fig. 3. Change in RER at varying exercise intensities.

Fatigue for prolonged exercise is due to hypoglycemia, muscle glycogen depletion, or dehydration. The first two reflect the important contribution of carbohydrates for exercise. *Fig. 4* clearly demonstrates that an elevated muscle glycogen level after 3 days of a high carbohydrate diet results in an enhanced exercise time to exhaustion. It also shows that a low muscle glycogen concentration results in an attenuated time to exhaustion.

Fig. 5 shows that **hypoglycemia** may be a cause of fatigue as can be seen at point A where the blood glucose level had fallen to around 3 mM. At this time the subject was given 100 grams of glucose and then was able to exercise for a further 40 minutes before fatigue ensued. At the second point of fatigue, blood glucose was above 4 mM and so hypoglycemia could not have been the cause.

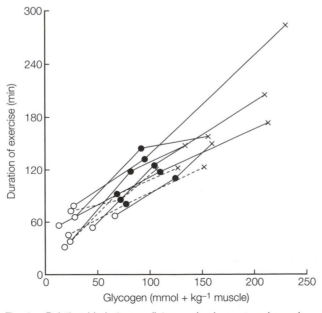

Fig. 4. Relationship between diet, muscle glycogen and exercise capacity.

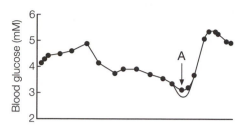

Fig. 5.   Blood glucose concentration over time during a test to fatigue.

Intermittent exercise

Many sports require the athlete to engage in repeated bouts of high-intensity activity with lower levels of activity. A game of soccer for example consists of 1% sprinting, 4% cruising, 32% jogging, 48% walking, and 15% standing still. In other words, the game of soccer draws approximately 95% of its total energy from aerobic sources and 5% from anaerobic sources. This means that the majority of the energy arises from carbohydrate and fat stores.

Use of match analysis techniques, by analyzing video footage of sports, leads to a greater understanding of the game demands and an awareness of the overall intensities and types of activities in that sport. The addition of heart rate recordings during sports, in conjunction with the match analysis, helps to confirm the levels and patterns of activity.

Fatigue during intermittent exercise is likely to result from hypoglycemia, muscle glycogen depletion, and dehydration as for prolonged exercise, although in addition any gradual and sustained increases in lactic acid concentrations during the intense activity phases may also be a factor. To illustrate the point about muscle glycogen being a possible factor in fatigue during intermittent activity, the findings from a study on soccer players can be seen in *Table 1*. It can be seen that of the soccer players who started the match with a low muscle glycogen content the distance covered in the second half of the game and the ability to sprint was severely diminished.

Table 1.   Relationship between resting muscle glycogen levels, distance covered, and % sprints during a soccer match.

	Normal glycogen	Low glycogen
Muscle glycogen before game	96 mM kg^{-1}	45 mM kg^{-1}
Muscle glycogen at half time	32	6
Muscle glycogen at end of game	9	0
Distance covered in 1st half	6100 m	5600 m
Distance covered in 2nd half	5900 m	4100 m
% time spent walking	27	50
% time spent sprinting	24	15

B6 RESPONSES TO TRAINING

Key Notes

Adaptation	When muscles undergo a period of training, their enzyme activity and structure become modified in response to the training. This is known as adaptation.
Muscle fiber type	There are essentially three types of muscle fibers, i.e. slow oxidative (SO or type I), fast glycolytic (FG or type II$_x$), and fast oxidative-glycolytic (FOG or type II$_a$). These respond to training, although the magnitude of the response for each fiber type depends on the training.
Endurance training	This type of training promotes the aerobic development of muscle fibers and is usually prolonged exercise undertaken at a relatively low exercise intensity. It may be continuous or intermittent.
Sprint training	This form of training enhances the anaerobic capacity and power generative capacity of muscles. Sprint training normally involves repeated efforts at a high exercise intensity.
Related topics	Adaptations to training (C5) The endocrine system (F2) Cardiovascular responses to Training for performance (Section H) training (E4)

Adaptation

Adaptation occurs in muscles as a consequence of repeated bouts of exercise over a period of time. The form of adaptation can be structural, in which case there is a modification of the **actin** and **myosin**, or functional, in which changes occur with respect to mitochondrial density and cytoplasmic enzyme activity.

Muscle fiber type

Muscle fibers generate the power and energy for work to be undertaken and are essentially of three types based on their contractile and metabolic properties. *Table 1* highlights the essential differences between the muscle fiber types. The **SO** fibers contain many mitochondria and have a high capillary network supplying blood to them. The net result is that they are the muscle fibers engaged in prolonged, lower levels of exercise. On the other hand, the **FG** fibers have fewer mitochondria and capillaries supplying blood, and consequently are the fibers employed in generating high levels of force.

Endurance training

Endurance training is that type of training in which the muscle fibers are recruited at lower levels of exercise intensity for a prolonged period of time (see Section H). The energy demands for this type of activity are derived mainly from aerobic sources and are likely to tax the SO and **FOG** fibers rather than the FG fibers. Consequently the adaptations that occur do so in these muscle fibers. The resulting changes are an improved ability to utilize oxygen due to increases in capillarization of the muscle fibers as well as increases in the

Table 1.   Characteristics of muscle fibers.

	SO (Type I)	FOG (Type II$_a$)	FG (Type II$_x$)
Contractile			
Speed of contraction	Slow	Fast	Fast
Force production	Low	Intermediate	High
Fatiguability	Low	Intermediate	High
Metabolic			
Glycogen stores	Low	High	High
Glycolytic enzyme activity	Low	High	High
Oxidative enzyme activity	High	Intermediate	Low

number of mitochondria in the cells. Since the mitochondria contain the enzymes for the TCA cycle and for β oxidation of fatty acids, there is an improvement in aerobic capacity. The net result of endurance training can be measured in increases in $\dot{V}O_{2max}$, a shift in the lactate threshold to the right, and faster time to undertake a set distance.

Sprint training   Sprint training involves repeated bouts of high-intensity efforts interspersed with appropriate recovery periods (see Section H). Since the bouts of activity are intense, the FG and FOG fibers are mainly recruited to generate the force. The type of energy source used for each sprint bout is dependent on the time i.e. 6 s or 45 s or 60 s intervals. The resultant effect is that the enzyme activity for that energy source is enhanced with training, i.e. repeat 5 s bouts are more likely to enhance the PCr system whereas repeat 30 s bouts are likely to promote improvements in the glycolytic energy system. Changes can thus be seen in the enzyme activities of creatine kinase, glycogen phosphorylase, and PFK in the FG and FOG fibers.

C1 MUSCLE STRUCTURE

Key Notes

Gross structure	Skeletal muscle under a microscope has well-defined striations, and with the exception of some facial muscles act across joints to produce movement.
Muscle fibers	A muscle is constructed from thousands of cylindrical elongated cells called muscle fibers.
Sarcolemma and sarcoplasm	Each muscle fiber is surrounded by a plasma membrane called the sarcolemma. This membrane acts to control active and passive transport into the cell. The fluid enclosed within the cell is the sarcoplasm. The sarcoplasm houses the cell nuclei, sarcoplasmic reticulum and the contractile apparatus.
Myofibrils	Each muscle fiber contains bundles of smaller myofibrils. There are approximately 2000 myofibrils in an adult muscle fiber.
Myofilaments	Each myofibril consists of a bundle of myofilaments. These filaments represent the contractile apparatus of the muscle and are either thin (actin) or thick (myosin) filaments.
Triad	A single transverse tubule and two terminal cisternae (sacs of the sarcoplasmic reticulum) form a triad. The triad aids rapid communication between the sarcolemma and the contractile apparatus.
Sarcomere	A myofibril is constructed from a series of sarcomeres added end to end. Each sarcomere contains interdigitating actin and myosin filaments separated from the next sarcomere by Z bands.
Related topics	Fiber types (C4) Cardiovascular structure (E1) Adaptations to training (C5)

Gross structure Skeletal muscle tissue is highly specialized to generate force and thus movement. The major function of muscle is to produce motion, to aid in the maintenance of posture, and to produce heat. In order to provide these functions muscle tissue can respond to stimuli, can conduct a wave of excitation, can modify its length and can regenerate in growth. These adaptabilities are referred to as the plasticity of muscle. A single muscle, as seen in *Fig. 1*, is constructed of a bunch of muscle fibers, and these fibers are linked together by a collagenous connective tissue. This connective tissue is networked into three layers: (a) the **endomysium** surrounds individual muscle fibers, (b) the **perimysium** collects bundles of fibers into fascicles, and (c) the **epimysium** provides a sheath around the entire muscle. This continuous fascia connects muscle fibers to tendons and consequently to the periosteum of bone, and the entire unit is known as the musculotendinous unit.

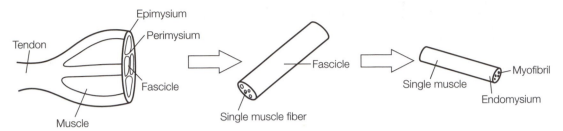

Fig 1.   The gross structure of skeletal muscle.

Muscle fibers

If skeletal muscle is viewed under a microscope it is seen to consist of thousands of elongated cylindrical cells called **muscle fibers** or **myofibers**. These fibers vary in length between 1–400 mm, and can be 10–100 μm in diameter.

Sarcolemma and sarcoplasm

Each muscle fiber is surrounded by a plasma membrane called the **sarcolemma**; this membrane allows for both passive and active transport into the cell and thus is an important factor in muscle excitability. The fluid enclosed within the cell by the sarcolemma is termed the **sarcoplasm**, and this contains fuel sources such as lipids and glycogen, organelles such as the cell nuclei and mitochondria, and the enzymes and contractile proteins required for muscle contraction.

Myofibrils

Also contained within the sarcoplasm are the **myofibrils**. These are cylindrical structures, approximately 1–2 μm in diameter that run longitudinally through the muscle fiber. Each muscle fiber can contain several hundred to several thousand myofibrils.

Myofilaments

The myofibrils consist of a bundle of smaller structures called **myofilaments**. These structures are approximately 6 nm (**thin filaments**) or 16 nm (**thick filaments**) in diameter, and represent the contractile apparatus of the muscle. The thin myofilaments are composed mostly of the protein **actin**, which is arranged in two strands intertwined into a helical structure. Within the groove of this helix structure sit two strands of the protein **tropomyosin**, upon which at regular intervals sits the protein **troponin**. The troponin complex includes three subunits: (a) **troponin I**, which binds to actin, (b) **troponin C**, which binds to calcium ions, and (c) **troponin T**, which binds to tropomyosin. The thick filaments are composed mostly of the protein **myosin**. Myosin has a two-chained helical tail, that at one end terminates in two large globular heads. These heads, during contraction, are referred to as **cross-bridges** and contain an ATP-binding site that is imperative for muscle contraction to occur.

Triad

The sarcoplasm also contains a hollow membranous system that is linked to the sarcolemma and assists in conducting neural commands through the muscle. This system includes the **sarcoplasmic reticulum**, the **terminal cisternae** and the **transverse tubules** (T tubules). The sarcoplasmic reticulum runs longitudinally along the fiber and surrounds the myofibril and at specific points it dilates into **lateral sacs**, or terminal cisternae. Running perpendicularly to the sarcoplasmic reticulum are transverse tubules that open to the outside of the fiber. A single transverse tubule plus two terminal cisternae, one on each side of the tubule, form a **triad**. The triad is extremely important in aiding the rapid communication between the sarcolemma and the contracile apparatus required for successful skeletal contraction.

Sarcomere

A myofibril is constructed from a series of **sarcomeres** added end to end, and it is these sarcomeres that give the muscle its striated appearance under the microscope. The structures of the sarcomere, seen in *Fig. 2*, are identified by how they refract light under the electron microscope. At rest a sarcomere is approximately 2.5 µm in length and is separated from its neighboring sarcomeres by narrow zones of dense material called **Z lines** or **Z bands**. The zone containing thick filaments and some interdigitating thin filaments is doubly refractive of light, or anisotropic, and is termed the **A** or **dark band**. The area between the A bands is singly refractive (isotropic) and is thus called the **I** or **light band**. This band contains mainly thin filaments. The area in the A band where there is no overlapping of thin filaments is called the **H zone**, in the center of which lies the **M line** or **M band**. The thin filaments are anchored to the Z lines and project in both directions, whilst the thick filaments attach to the M band. During muscle contraction the myosin cross-bridges pull on the actin filaments so that they slide inwards towards the H zone. The sarcomere shortens as the Z lines move towards each other, but the length of the myofilaments does not change. As the thin filaments meet at the center of the sarcomere the H zone narrows or disappears. The shortening of muscle fibers by the sliding of myofilaments is called the **sliding-filament theory**.

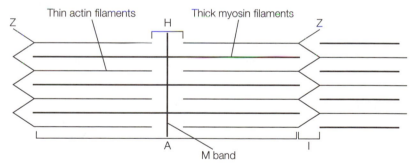

Fig 2. Structure of the sarcomere.

C2 MOTOR-NEURAL CONTROL OF CONTRACTION AND RELAXATION

Key Notes

The nervous system	The nervous system is the chief homeostatic operator of the body. Consisting of the central and peripheral systems its function is to sense, interpret and respond to alterations within the body.
Neuron	The neuron is the cell responsible for conducting nervous impulses. The cell consists of a large cell body containing the nucleus of the cell, cytoplasmic processes called dendrites responsible for conducting impulses into the cell, and a long process called an axon responsible for conducting impulses away from the cell body.
Nerve impulse	The nerve impulse is often referred to as the electrical message sent along the neuron. The impulse is in fact an alteration in the difference in ionic charge across the membrane of the neuron. At rest this charge is negative inside the cell and positive outside and is termed the resting membrane potential. The impulse is carried down a neuron by reversing this charge, a process termed depolarization. The charge across the membrane at rest or during depolarization depends upon the concentration of charged ions each side of the membrane. The major ions of interest are sodium (Na^+) and potassium (K^+).
Synapse	The nerve impulse is passed from neuron to neuron, or from neuron to muscle or gland. The space between the end of the neuron and its target is termed the synapse. A nervous impulse is carried across the synaptic cleft by a chemical neurotransmitter. This neurotransmitter stimulates the postsynaptic membrane in order for the impulse to be conducted further.
Neuromuscular junction	A neuron stimulating muscle tissue synapses with the sarcolemma of a muscle fiber. This synapse is termed the neuromuscular junction.
Motor unit	The neuron stimulating a muscle tissue, and the muscle fibers it innervates are collectively termed the motor unit.
Excitation–contraction coupling	The nervous impulse arriving at a muscle is transmitted down the t tubules to the triads where a series of chemical events causes a muscular contraction to occur. This conversion of an impulse into the attachment of cross-bridges is termed excitation–contraction coupling.
Cross-bridge cycle	The attachment of the myosin head to the actin filament and its rotation causing the filaments to slide across each other is termed the cross-bridge cycle. This cycle is powered by ATP and dependent upon the release of calcium ions (Ca^{2+}) from the sarcoplasmic reticulum.

Proprioception	Neural control of skeletal contraction and thus movement of the body and limbs is dependent upon the CNS receiving feedback about where the individual body parts are in relation to each other. This feedback to the CNS is provided from receptors in the periphery termed proprioceptors.
Related topics	Energy sources and exercise (B1) Adaptations to training (C5)

Cardiovascular function and control (E2)

Neural system (F1)

The nervous system

The nervous system is the body's chief homeostatic operator. Its function is: (i) sensory (to detect changes in the body and the environment); (ii) integrative (to interpret these changes); and (iii) motor (to respond to these changes with movement or endocrine secretions). The nervous system can be divided into two major components, **the central nervous system** and the **peripheral nervous system**. The central nervous sysytem (CNS) consists of the brain and the spinal cord, and essentially all receptors must relay sensations to the CNS for responses to occur. The nerves, and nerve processes, that connect the CNS to the muscles, glands and receptors make up the peripheral nervous system (PNS). The PNS is subdivided into an **afferent** and an **efferent** system, whereby afferent nerves carry impulses towards the CNS, and the efferent nerves carry impulses from the CNS to the periphery. The efferent system is further subdivided into a **somatic** and **autonomic** nervous system. The somatic nervous system is made up of efferent nerves carrying impulses to skeletal muscles, whilst the autonomic nervous system consists of efferent nerves carrying impulses to cardiac and smooth muscle tissue, and glands. The somatic system is under conscious control and thus produces voluntary movement. The autonomic system produces involuntary movement or responses, and is split into **sympathetic** and **parasympathetic** divisions. In general one division serves to increase function, whereas the other division serves to decrease function. The sympathetic and parasympathetic divisions of the nervous system are explained in more detail in Section F. The structure of the nervous system is reviewed in *Fig. 1*.

Neuron

The two principal cells found in the nervous system are the neuroglia and the neuron. The **neuroglia**, or **glial cells**, form a supporting and binding structure for the neurons, whilst some specialist cells (**Schwann cells**) produce the phospholipid, **myelin**, to cover the nerve fiber and increase the speed of nerve impulse conduction. The **neuron** is the cell responsible for conducting nerve impulses and consists of three distinctly recognizable structures (*Fig. 2*). The **cell body**, **soma** or **perikaryon** contains the cell nucleus and nucleolus surrounded by a granular cytoplasm. This cytoplasm contains the functional apparatus of the cell, i.e. mitochondria, lysosomes, golgi apparatus and nissl bodies. The neuron has two cytoplasmic processes, the dendrites and the axon. The **dendrites** are branched, thick extensions of the cytoplasm whose purpose is to conduct nerve impulses towards the cell body. The **axon** is a single, long process that originates from the cell body from the **axon hillock**. This process conducts nerve impulses away from the cell body to another neuron, or muscle or gland. The cytoplasm of the axon is called the **axoplasm** and is surrounded by a plasma membrane called the **axolemma**. Axons vary in length from between 1 mm to over 1 m. Along its length there are branches called

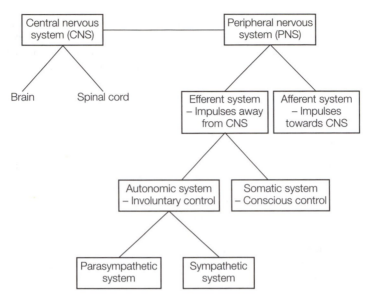

Fig. 1.   Schematic of the structure of the nervous system.

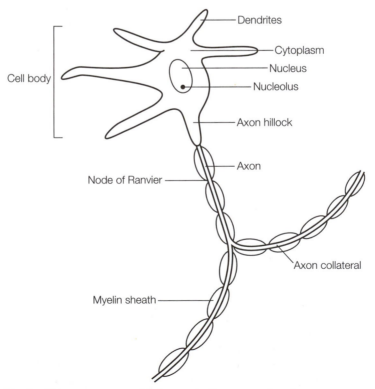

Fig. 2.   The neuron responsible for conducting nervous impulses.

axon collaterals which terminate in fine filaments called axon terminals. The ends of these terminals are bulbous and are termed synaptic end bulbs. These bulbs contain synaptic vesicles storing neurotransmitters, and are imperative for transport of a nerve impulse from one neuron to another, or to muscle or glandular tissue.

Nerve impulse

The function of neurons is essentially to conduct electrical messages called nerve impulses. The ability to generate and conduct nerve impulses is a function of the excitable nature of the cell membrane. The membrane is surrounded by a different concentration of ions in the intra- and extracellular fluids. The major ions of interest are sodium (Na^+) and potassium (K^+), and in a resting neuron the K^+ concentration inside the cell is approximately 30× greater than outside, whilst the Na^+ concentration is around 14× greater. These differing concentrations cause a charge difference across the membrane known at rest as the resting membrane potential (60–90 mV). The net positive charge outside, and negative charge inside the membrane, or polarization of the membrane, is maintained by the pumping action of a membrane-bound protein known as the sodium–potassium pump. This pump keeps Na^+ in the extracellular fluid and K^+ in the intracellular fluid by actively (requires ATP) transporting Na^+ out of the cell. K^+ remains inside the cell by the action of the pump and by its attraction to negatively charged anions (amino acids and proteins) that are too large to cross the membrane. Any condition that serves to alter this resting membrane potential is referred to as a stimulus, and might be induced chemically (neurotransmitters) or electrically (movement of ions across the membrane).

If a stimulus is applied to a polarized membrane, the membrane's permeability to Na^+ ions increases. This leads to an influx of Na^+ into the cell via Na^+ channels, and thus a change in the membrane potential. The potential inside the cell (usually –70 mV) rises towards 0, and then rises to a positive 30 mV. This process is termed depolarization and is dependent upon voltage-sensitive (-gated) channels in the cell membrane. These channels are composed of proteins within the membrane that can alter shape in response to the electrical potential (or voltage) of the membrane. In other words they act as a gate for movement of Na^+ and K^+ across the cell membrane.

The action of depolarization initiates a nerve impulse or nerve action potential. An action potential initiated at one point along the nerve fiber usually excites (depolarizes) the adjacent portion of the fiber. In other words the impulse or action potential is passed along the fiber (propagated). This propagation relies on the voltage-sensitive ion channels to allow for the membrane potential to change at each point along the fiber. Depolarization takes approximately 0.5 ms, and then the resting potential is restored (repolarization). Repolarization, or restoring the membrane potential to its resting state, is dependent upon voltage-sensitive K^+ channels. As the Na^+ channels are becoming inactivated the K^+ channels open and K^+ diffuses out of the cell leaving the inner surface of the membrane negatively charged once again. The delay in the activation of the K^+ channels leads to a slight hyperpolarization of the membrane. Eventually the ions are restored to their original site via the sodium–potassium pump. During repolarization the neuron cannot conduct another nerve impulse and this timespan is termed the refractory period. This period is approximately 0.4 ms in large nerve fibers and 4 ms in small fibers. This whole process is demonstrated in *Fig. 3*.

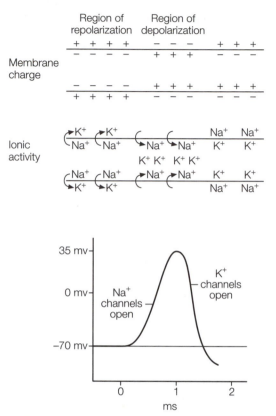

Fig. 3.   The action potential.

A stimulus strong enough to conduct a nerve impulse is termed a **threshold stimulus** as the neuron has reached its threshold of stimulation. Any stimulus weaker than the threshold is termed a **subthreshold stimulus**. A single nerve cell that depolarizes transmits the action potential along the nerve fiber according to the **all-or-none principle**. That is, if a stimulus is strong enough to generate an action potential, the impulse will be conducted along the entire neuron.

The speed at which an impulse travels varies up to 120 m sec^{-1} with the diameter of the axon. This is because the surface area of large axons is insulated with the phospholipid myelin, which inhibits ionic movement. The myelin sheath is interrupted at intervals with non-insulated axolemma called **nodes of Ranvier**. At these nodes depolarization can occur, and so the action potential is propagated by jumping from node to node. This is termed **saltatory conduction** and in fact involves the movement of ions through the extracellular fluids and axoplasm. Axons insulated with myelin conduct impulses at a faster rate than unmyelinated fibers.

Synapse

The nerve impulse is not only conducted along each neuron, but also from neuron to neuron, and from neuron to muscle or gland. The junction between two neurons is termed a **synapse**, and this is essential for homeostasis as it possesses the ability to both conduct and inhibit the action potential. Two neurons are separated by a small space filled with extracellular fluid known as the **synaptic cleft**. The neuron before the cleft is termed the **presynaptic** neuron, and the **postsynaptic neuron** is situated after the cleft. The presynaptic neuron ends in a bulbous structure called the **synaptic end bulb**, and can synapse with dendrites (**axodendritic**), the cell body (**axosomatic**) or the axon hillock (**axoaxonic**) of the postsynaptic neuron. The presynaptic neuron may synapse with a number of postsynaptic neurons (**divergence**) or merely with a single postsynaptic neuron (**convergence**).

The nerve impulse is passed across the synapse either electrically or chemically. In the electrical transmission the impulse passes through protein tubular structures called **gap junctions**. This is specifically the case in smooth muscle and cardiac tissue. At a chemical synapse the neuron secretes a **neurotransmitter** into the cleft that acts upon the receptors on the postsynaptic neuron, or muscle or gland tissue. Neurotransmitters are produced by the neuron, transported to the synaptic end bulb and stored in sacs called **synaptic vescicles**. Once an action potential reaches the end bulb the vesicles are attracted to the membrane and the neurotransmitter molecules are released into the synaptic cleft. The impulse can only be conducted in one direction and each neuron will contain only one type of neurotransmitter. The neurotransmitter will then diffuse across the synaptic cleft and bind with its own receptors on the postsynaptic membrane. If sufficient neurotransmitter binds to the receptors Na^+ channels on the postsynaptic membrane open and the membrane depolarizes. In this way the impulse is propagated across the synaptic cleft. Increases in the postsynaptic membrane potential are termed **excitory postsynaptic potentials**. Some synapses act as inhibitory synapses whereby the postsynaptic membrane is made more negative by the opening of K^+ channels. Excitory postsynaptic potentials can occur by two means: (i) **temporal summation** or the summing of several action potentials from one presynaptic neuron over a period of time, or (ii) **spatial summation** or the summing of several action potentials from a number of presynaptic neurons. Synaptic transmission is demonstrated in *Fig. 4.*

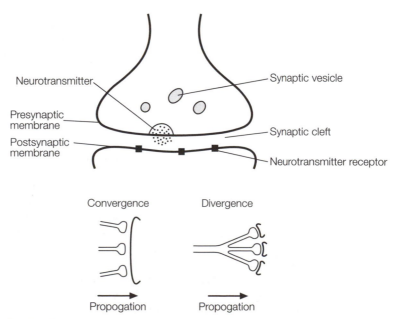

Fig. 4.   The synapse and synaptic transmission.

Neuromuscular junction

A neuron stimulating muscle tissue is a **motor neuron** whose axon terminals come into close contact with the sarcolemma of a muscle fiber. The area of the sarcolemma adjacent to the axon terminal is termed the **motor end plate**, and it is this end plate and the presynaptic axon terminal that form the **neuromuscular junction**. The neurotransmitter released at neuromuscular junctions in skeletal muscle is **acetylcholine** (Ach). This Ach diffuses across the synaptic cleft and combines with receptor sites on the sarcolemma of the muscle fiber. Once the action potential has been propagated the Ach in the cleft is degraded into acetate and choline, a reaction catalyzed by **acetylcholinesterase**. The depolarization of the motor end plate propagates the action potential or nerve impulse into the muscle via the t-tubules.

Excitation–contraction coupling

The transmission of the nerve impulse across the neuromuscular junction and down the t tubules leads to a series of chemical events that causes the muscle to contract via the **sliding filament theory**. The conversion of a nervous impulse into the interaction of actin and myosin (the cross-bridge cycle) is known as **excitation–contraction coupling**. The neuron, branches of its axon and the muscle fibers that it innervates are known collectively as the **motor unit**.

Cross-bridge cycle

Under resting conditions actin and myosin are prevented from interacting by the action of the proteins **troponin** and **tropomyosin**. When an action potential is propagated down the t-tubules it increases the permeability of the membrane of the terminal cisternae to **calcium** (Ca^{2+}). Calcium is imperative for muscle contraction and is stored in the sarcoplasmic reticulum. This increase in permeability leads to Ca^{2+} diffusing into the sarcoplasm where it binds to troponin C. This binding causes a structural rotation of the troponin–tropomyosin complex uncovering the mysosin binding site on the actin filament. The head of the myosin filament now attaches to the binding site, ATP is hydrolyzed to ADP + Pi (catalyzed by ATPase stored in the myosin head) and the energy released is used to rotate the myosin head (see Section B for energetics). As the myosin head is now attached to the actin filament this rotation causes the filaments to slide across each other. The filaments then detach and the myosin head returns to its original position. Once contraction is finished Ca^{2+} is returned to the sarcoplasmic reticulum via a Ca^{2+} **pump**, and the myosin binding site on the actin filament is hidden once more. This process is reviewed in *Fig. 5*.

Proprioception

Neural control of skeletal contraction and thus movement of the body and limbs is dependent upon the CNS receiving feedback about where the individual body parts are in relation to each other. This feedback to the CNS is provided from receptors in the periphery termed **proprioceptors**. Proprioceptors can be free nerve endings, Golgi-type receptors or pancinian corpuscles. Free nerve endings are sensitive to touch and pressure and as such are stimulated at the beginning of movement, becoming less sensitive as movement continues. Golgi-type receptors are position receptors located in ligaments around joints, whilst pancinian corpuscles are located in the tissue surrounding joints and detect the rate of joint rotation. As a whole these receptors provide the CNS with feedback on the orientation and speed of limb movements.

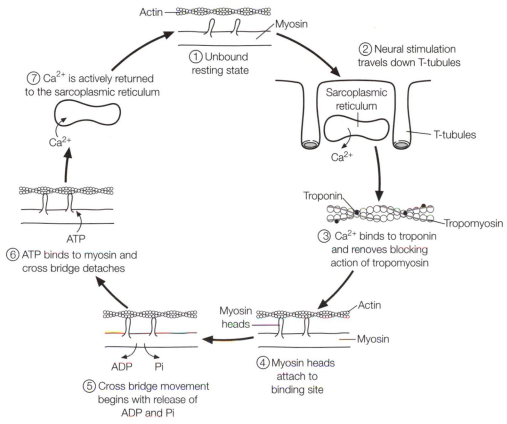

Fig. 5. The cross bridge cycle.

The skeletal muscle contains its own receptors to aid in fine muscle control, these are chemoreceptors, muscle spindles and Golgi tendon organs. **Chemoreceptors** are free nerve endings that provide feedback to the CNS regarding pH, K^+ concentration and oxygen and carbon dioxide tension in the muscle. The **muscle spindle** provides the CNS with feedback regarding relative muscle length, both in static and dynamic contractions, whilst the **Golgi tendon organs** are located within the muscle tendon to monitor muscle tension. Muscles requiring fine muscular control, such as those in the hand, have a greater number of muscle spindles than others. The spindles lie parallel with the muscle fibers and have neurons that synapse at the spinal cord. This allows the receptors to feed back to the CNS and receive rapid responses without involving higher brain centers. This type of feedback arc is termed a reflex response, a good example of which is the **mytatic** or **stretch reflex** explained in *Fig. 6.*

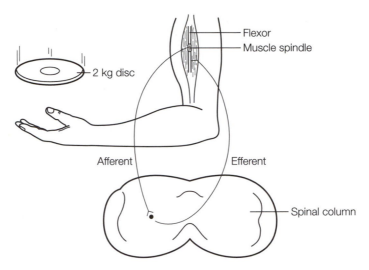

Fig. 6.   The stretch reflex. A disc dropped onto the hand causes a stretch of the flexor muscle of the upper arm. This stretch is detected by the muscle spindle which reflexly aids the correction of the stretch.

C3 FORCE–VELOCITY AND LENGTH–TENSION CHARACTERISTICS

Key Notes

Twitch and tetanus	In the laboratory a single muscle fiber can be examined. Low-frequency stimuli applied to the fiber produce a single contraction called a twitch. Higher frequencies cause the twitches to summate into a sustained contraction called tetanus.
Recruitment	The force generated by a muscle or group of muscles is influenced by the numbers and types of motor units recruited.
Discharge	The force produced by a muscular contraction is also influenced by the rate and pattern of action potential discharge.
Muscle mechanics	The force generated by a muscle is dependent upon its length and the velocity of contraction. The force–length relationship determines the number of cross-bridges likely to be attached. The force–velocity relationship dictates that force decreases as the velocity of contraction increases.
Related topics	Motor-neural control of contraction and relaxation (C2) Neural system (F1)

Twitch and tetanus

A single stimulus to a muscle fiber will result in a brief contraction followed by an immediate relaxation and is called a **muscle twitch**. Additionally, a series of low-frequency stimuli will result in a series of twitches. As the frequency of stimuli increases the muscle does not have time to relax between them and thus the force produced beomes additive via a **summation** of twitches. Eventually individual contractions blend together in order to produce a single contraction called **tetanus** (*Fig. 1*). Tetanus will continue until the stimuli is removed, or the muscle fatigues. In normal movement contractions are tetanic and the phasing of muscle tetany and relaxation across the body leads to controlled and smooth muscular motion.

Recruitment

The recruitment of additional motor units acts to increase the force that a muscle can exert. The order of **motor unit recruitment** is influenced by size; that is, the unit with the smallest motor neuron is recruited first and that with the largest last. Recruitment is also influenced by membrane sensitivity and the number of synapses on a specific motor neuron. This means that recruitment is orderly and predetermined and that as the force required for a task increases the recruitment of larger motor units will progressively increase.

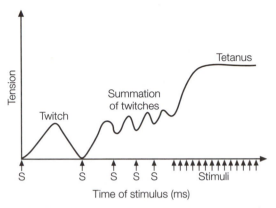

Fig. 1.   Muscle twitch and tetanus as the frequency of stimulus increases.

Discharge

The force produced by a muscular contraction is also influenced by the rate and pattern of **action potential discharge**. As force increases the rate of discharge of action potentials increases, such that two action potentials discharged very rapidly may summate and cause a greater exertion of force.

Muscle mechanics

As muscular contraction depends on the cross-bridge cycle, the development of force is also dependent upon cross-bridge formation and detachment. The number of cross-bridges able to form is dependent upon the muscle length, and thus muscle length at the time of contraction also influences the ability to produce force. A muscle fiber produces its maximum tension when there is maximum overlap of the actin and myosin filaments (maximum number of cross-bridges in formation), and this is the **optimal length**. As a fiber is stretched fewer cross-bridges can form and thus the tension that can be developed is decreased. At lengths less than the optimal the myosin filaments crumple as they run into the Z lines, and thus the force of contraction decreases. This relationship between force and fiber length is termed **the length–tension or force–length relationship**. *Fig.* 2 displays this relationship for isometric contraction (force is generated but the fibers do not change length).

The force generated by muscle is also dependent upon the velocity of fiber shortening or lengthening. This is known as **the force–velocity relationship**. A contraction causing the muscle fiber to shorten is known as a **concentric**

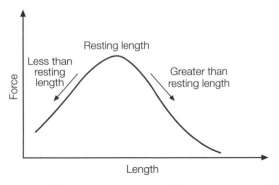

Fig. 2.   Relationship between muscle length and muscle force for an isometric contraction.

contraction. As the velocity of shortening increases the force produced decreases (*Fig. 3*). The force at zero velocity of shortening (**isometric force**) is termed Po, and many textbooks will refer to forces (P) produced at differing velocities as a fraction of Po. When the muscle is shortening P is less than Po because there are fewer cross-bridges attached during shortening, and each cross-bridge exerts less force than in the isometric state. The faster the shortening occurs the less tension is generated in the myosin filament, and in some cases the myosin head does not have time to detach and actually resists movement. If the force generated by cross-bridges resisting movement equals that being generated by those cross-bridges acting normally no force will be developed. The velocity at which this occurs is termed V_{max}.

Muscle contractions occurring as the muscle elongates are termed **eccentric contractions.** The force generated by muscle during this type of contraction is greater than Po and varies with velocity. As the velocity of stretch increases the number of attached cross-bridges decreases, however those attached sustain more force. *Fig. 3* displays this classic relationship between force and velocity for isometric, concentric and eccentric contractions.

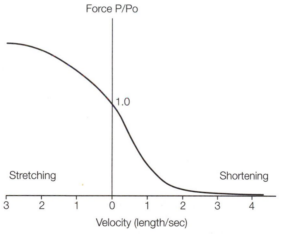

Fig. 3.   The relationship between the velocity of muscle contraction and muscle force. Po is the isometric force. P is the force produced at a given velocity.

C4 FIBER TYPES

Key Notes

Muscle fiber type	Muscle fibers can be classified as either fast or slow twitch dependent upon their structure and chemical make-up.
Fast-twitch fibers	Fast-twitch fibers (type II) are larger than slow twitch, have a high concentration of myosin ATPase and thus rely on short-term glycolytic energy transfer. This makes them suited for anaerobic activities such as sprinting, weight-lifting and jumping. Fast-twitch fibers can be further classified into type IIa and type IIb or IIx fibers. Type IIb/IIx are the fast-twitch glycolytic fibers, wheras the fast-twitch type IIa are oxidative glycolytic fibers suited to middle-distance events.
Slow-twitch fibers	Slow-twitch fibers (type I) are fatigue-resistant and thus suited to longer-term, aerobic activities. Their high concentration of myoglobin and mitochondria allows reliance upon oxidative energy transfer.
Fiber type distribution	Different muscles possess differing concentrations of fast- and slow-twitch fibers dependent upon their function. Fiber type distribution is genetically determined, however training can optimize the function of each fiber type or cause transition of fiber types.
Related topics	Responses to training (B6)　　　　　Adaptations to training (C5)

Muscle fiber type　Each motor unit contains fibers of the same type. Historically, muscle fibers have been identified as **red** or **white** fibers, where red fibers were suited to long-term slow muscle contractions (**slow-twitch**) and white fibers to high-speed contractions (**fast-twitch**). As science has advanced and the techniques for examining muscle fibers have allowed analysis at the cellular level, the structure and function of differing fiber types has been further elucidated (see also Section B). The three major muscle fiber types seen in human skeletal muscle are:

- slow-twitch oxidative (type I)
- fast-twitch oxidative-glycolytic (type IIa), and
- fast-twitch glycolytic (type IIb). More recent work has identified a fourth fiber type similar to IIb labeled type IIx. This chapter will refer to both of these fiber types as type IIb/x.

Fast-twitch fibers　**Fast-twitch fibers** are generally larger than slow-twitch and have the ability to generate energy rapidly for quick and explosive actions. This ability is related to a capability for rapid transmission of action potentials, a high activity level of myosin ATPase and a highly developed sarcoplasmic reticulum. This means that the fiber can rely on short-term glycolytic energy systems (see Section B for more detail) and is thus activated in anaerobic activities such as sprinting, lifting, jumping, etc. Indeed, the fast-twitch fiber can develop tension 3–5 times faster than the slow-twitch fiber with the fastest contractions produced by the type IIb/x fibers. The type IIa fibers are intermediate fibers in that they can contract quickly but also have the capacity to utilize energy from oxidative processes. Thus the resistance to fatigue is greater in type IIa fibers than in type IIb/x. A muscle containing a greater percentage of fast-twitch fibers can produce more power at any given velocity of contraction, and can contract at a greater velocity at any given force.

Slow-twitch fibers   **Slow-twitch fibers** are fatigue-resistant and thus suited to longer-term, aerobic activities. They have greater myoglobin and mitochondria levels than fast-twitch fibers, and are thus reliant upon aerobic metabolism to function. They contract at a much slower speed than fast-twitch fibers and have a low glycolytic capacity. The major differences between the fiber types are displayed in *Table 1*.

Table 1.   Characteristics of skeletal muscle fiber types

Characteristic	Type IIb/x	Type IIa	Type I
Force production	High	High	Slow
Predominant energy system	Anaerobic	Combination	Aerobic
Fatigue resistance	Low	High/Moderate	High
Myosin ATPase activity	High	High	Low
Oxidative enzyme activity	Low	High	High
Phosphocreatine stores	High	High	Low
Glycogen stores	High	High	Low
Mitochondrial density	Low	High	High
Fiber diameter	Large	Large	Small
Capillary density	Low	Medium	High
Speed of shortening	High	Intermediate	Low
Nerve conduction velocity	High	High	Low

Fiber type distribution   The average composition of muscles in the legs and arms is 45–55% slow-twitch fibers and equal numbers of type IIa and type IIb/x fast-twitch fibers. Other muscles may vary considerably in their fiber distribution depending upon their function. Postural muscles for example tend to have a greater percentage of slow-twitch, fatigue-resistant fibers, to suit their everyday function. It is generally accepted that fiber type distribution is genetically determined and does not differ between men and women, although men do tend to have larger muscle fibers than women. Endurance athletes have been seen to possess large percentages of slow-twitch fibers in the muscles generally used in their sport (90–95% in the gastrocnemius of cross-country runners and skiers). Weight-lifters, jumpers and sprinters have a greater percentage of fast-twitch fibers, whilst middle-distance athletes possess equal numbers of both fiber types. Muscle fibers, however, are capable of adapting their molecular composition in response to training stimuli. Training can thus result in a transition of one type of fiber into another (see Sections C5 and H4).

C5 ADAPTATIONS TO TRAINING

Key Notes

Type of training

Muscles respond to endurance, sprint and resistance training. This chapter will concentrate on resistance training.

Absolute or relative muscle strength

Muscle strength can be measured as the absolute amount of force produced or as the force produced relative to either body mass or lean tissue mass. This is important when comparing individuals of different sizes.

Muscle hypertrophy

Resistance training results in an increase in the size of muscle fibers via an increased synthesis of protein filaments. The increase in muscle size is termed hypertrophy.

Muscle hyperplasia

Resistance training in animals results in an increase in the number of fibers in addition to hypertrophy; whether this occurs in humans is currently debatable.

Fiber type composition

Resistance training results in transformation of type IIb/x fibers to type IIa fibers. The proportion of type I fibers in a muscle is not affected by resistance training. Endurance training results in fast to slow fiber transitions.

Connective tissue

The connective tissue sheath surrounding the muscle and muscle fibers increases in size as a consequence of training. Ligaments and tendons also increase in strength following training.

Motor-neural adaptations

Increases in the synchronization and efficiency of motor unit recruitment following training result in the major adaptations seen in muscular strength following resistance training programs.

Muscle soreness

Muscle soreness may occur after an acute training session, especially resistance training. The soreness manifests itself as stiffness, tenderness, inflammation and/or pain, and as it occurs between 24–48 hours following the training session is referred to as delayed-onset muscle soreness (DOMS).

Muscle atrophy and detraining

Detraining, or the cessation of physical activity following a training period, is associated with a decrease in muscular function. This decrease is also associated with sedentary living, space flight, long-term bed rest, limb casting and aging. Detraining results in a decreased muscle size termed atrophy or sarcopenia.

Related topics

Responses to training (B6)
Muscle structure (C1)
Fiber types (C4)
The endocrine system (F2)

Training for strength and
  power (H4)
Guidelines for exercise prescription
  (L2)

Types of training Muscles respond to endurance, sprint and resistance training. Endurance and high-intensity training are discussed in Sections B and H. This chapter will concentrate on responses to resistance training.

Absolute or relative muscle strength Muscle strength can be recorded in either absolute or relative terms. **Absolute muscle strength** refers to the absolute force produced, or weight lifted in a strength test. So an individual may be able to bench press 80 kg in one repetition, or produce a maximal knee extension of 190 Nm on an isokinetic dynamometer. This will probably mean (although not always) that larger people are stronger than smaller people, and that men are stronger than women. In reality it means that the individuals producing the bigger forces are in fact merely larger than their counterparts. In order to really compare strength in these individuals **relative muscle strength** should be compared. Muscle strength can be reported relative to body mass, lean tissue mass, muscle cross-sectional area or any other variable likely to influence performance. In this way strength per kilogram body mass, for example, can be compared between two individuals. Previous studies have indicated that if leg press strength is compared between males and females per kilogram of lean tissue mass, females are as strong, if not stronger than males.

Muscle hypertrophy Skeletal muscle growth, or **hypertrophy**, is associated with strength gains following resistance training. The increase in muscle size is from enlargement of muscle fibers. This occurs in both fast- and slow-twitch fibers, but to a greater extent in fast-twitch. The enlargement of the fibers is caused by an increase in the synthesis of protein filaments that constitute the actin and myosin filaments, and the number of sarcomeres. Muscle hypertrophy is an anabolic process and this is enhanced by increases in the concentration of testosterone, growth hormone and insulin-like growth factor I (IGF-I) following a resistance training regime (see Section F2). The catecholamines and cortisol act as catabolic hormones and the ratio of the anabolic and catabolic hormones is very much dependent upon fitness, nutritional status and type of resistance training undertaken.

Muscle hyperplasia Muscle **hyperplasia** refers to an increase in the number of muscle fibers, in addition to hypertrophy, as a result of training. Hyperplasia has been identified in animal studies but is questioned in human studies. If hyperplasia does occur in humans it may only occur at extreme levels of training when the fibers reach their upper limit in cell size.

Fiber type composition The **fiber type composition** of a muscle is genetically determined but training does appear to be able to alter this predetermined composition. Indeed a rapid effect of resistance training is a decreased percentage of type IIb/x fibers, with a concomitant increase in the percentage of type IIa fibers. This compositional change is via transformation of type IIb/x to type IIa fibers, a transformation that reverses in detraining. The proportion of type I fibers is not affected by resistance training. Endurance training has been seen to transform fast fibers to slow, but most training effects are in the transition of type II fiber isoforms.

Connective tissue The epimysium, perimysium and endomysium also have the capacity to adapt to resistance training. These **connective tissue sheaths** increase in size at the same rate as muscle tissue and act primarily as a support system for the muscle.

In addition the ligaments and tendons increase in strength providing a supporting framework for the muscle.

Certain metabolic **enzymes** have been seen to increase following resistance exercise training. **Creatine phosphokinase,** the enzyme associated with deriving energy from the ATP–PCr system (see Section B) is increased following isokinetic and isometric training using contractions greater than 6 seconds in duration. The enzyme is unaffected by isotonic resistance training, and none of the enzymes associated with the glycolytic, short-term energy systems are increased with resistance training. The enzymes associated with aerobic energy transfer have also been seen to elevate following isometric and isokinetic training, especially in type IIa fibers. On the whole, enzymatic alterations associated with resistance training are mostly related to the ATP–PCr system. For aerobic training changes see Section B.

Oxidative metabolism is supported by an increase in the capillary density of the muscle. **Capillarization** following resistance training depends upon the intensity and volume of the training program. High-intensity, low-volume strength training, for example power lifting, tends to decrease capillary density, whilst low-intensity, high-volume strength training increases capillary density. Thus, high-volume training may enhance performance of low-intensity tasks by increasing the blood supply to tissues.

Motor-neural adaptations

The increase in muscle strength seen following resistance training is only partly due to an increased muscle size. **Motor-neural adaptations** are thought to play the major part of the increase in strength, especially in the elderly. Certainly the initial changes in muscle strength are due to neural adaptations. Furthermore, following strength training on one arm, whilst there is hypertrophy in the trained arm there are gains in strength in both arms. The gains in the untrained arm are due to motor-neural adaptations. These adaptations include an increased size and complexity of the neuromuscular junction, an increased synchronization of motor units and a more efficient recruitment order of muscle fibers. In this way the amount of muscle required to be activated to produce a given force is less after training than prior to training.

Muscle soreness

Muscle soreness may occur after an acute training session, especially resistance training. The soreness manifests itself as stiffness, tenderness, inflammation and/or pain, and as it occurs between 24–48 hours following the training session is referred to as **delayed-onset muscle soreness** (DOMS). The exact mechanism for DOMS is not fully understood but it appears that acute exercise, especially eccentric contractions, causes damage to the ultrastructure, potentially the Z-lines, of the muscle cell. Plasma creatine kinase is used as a marker for muscle damage.

Muscle atrophy and detraining

Detraining, or the cessation of physical activity following a training period, is associated with a decrease in muscular function. This decrease is also associated with sedentary living, space flight, long-term bed rest, limb casting and aging. Detraining results in a decreased muscle size termed **atrophy** or **sarcopenia,** as a result of reductions in myosin and actin via a decreased fiber number and/or fiber size. The composition of fiber type tends to transform to its pretraining composition, whilst metabolically the enzymes associated with aerobic activity, and the mitochondrial content of the muscle are decreased. In addition the maximal firing rate of neurons, and the ability to recruit high-threshold motor neurons decreases following detraining. The significance of these physiological alterations is that both muscle strength and endurance are compromised, and as the major muscles affected tend to be those involved in weight bearing and postural control, movement is also hampered. Sarcopenia seen in aging may be more related to inactivity than with the aging process.

D1 PULMONARY STRUCTURE AND VOLUMES

Key Notes

Overview	The pulmonary system provides the first step in the 'oxygen cascade' which sees oxygen move from the air we breathe to the mitochondria of human cells. Other roles for the pulmonary system include the regulation of carbon dioxide levels in the body and to help the maintenance of acid–base status.
The lungs	The lungs, situated in the thorax, are the organs of pulmonary function. The lungs contain a number of structures for the delivery of oxygen from the air we breathe to the pulmonary circulation. Both the left and right lungs contribute to airway movement and gas diffusion.
The major airways	Major airways include the mouth and nasal cavities as well as the pharynx, larynx, trachea, bronchi and bronchioles. These airways serve as the major conduits through which air passes on its way from the external environment to the alveoli.
The alveoli and pulmonary capillaries	The diffusion of oxygen into the body occurs at the interface between the alveoli and the pulmonary capillaries. In reverse, carbon dioxide diffuses from the pulmonary capillaries into the alveoli.
Muscles associated with inspiration and expiration	The muscles associated with pulmonary function provide active support for both inspiration and expiration. The most important muscle is the diaphragm that contracts during inspiration to lower the floor of the thorax and raise the ribcage.
Ventilation and lung volumes	Ventilation is the movement of air in and out of the lungs. Movement of air into the lungs occurs in inspiration and out of the lungs in expiration. Ventilation will change in many circumstances and this will be reflected in changes in a range of lung volumes.
Related topics	Muscle structure (C1) Pulmonary function and control (D2) Motor-neural control of contraction Cardiovascular structure (E1) and relaxation (C2) The neural system (F1)

Overview

We have seen previously (see Section C) that ATP is produced via a number of metabolic pathways. Whilst not all **ATP** generated in the skeletal muscle is derived from the full oxidative breakdown of fats and carbohydrates it is clear that all forms of physical activity will promote oxidative metabolism in the muscles. To meet this demand the body initiates and controls (see Section F) a highly complex and integrated multi-organ response to deliver oxygen to cells within the body (e.g. muscle cells). The pulmonary system (or pulmonary respiration) is the first link in this process of oxygen delivery to the muscle tissue. This integral step in the 'oxygen cascade' begins with the transport of oxygen in the air we breathe to the alveoli in the lung. From here the oxygen diffuses into the pulmonary capillary circulation. As well as playing a vital role in the delivery of oxygen to the contracting muscles, the pulmonary system also serves to expel and thus regulate the levels of carbon dioxide within the body. As a by-product of tissue metabolism it is important to expel carbon dioxide in line with rates of production. This also forms part of the other major role for the pulmonary system, which is to help regulate the body's acid–base status.

The lungs

There are two lungs (left and right) that sit in a closed compartment called the **thorax** (**thoracic cavity**). The lungs are surrounded to the lateral, anterior, posterior and superior sides by the rib cage. The **diaphragm,** a band of muscle tissue, provides the inferior border of the thorax. A double-layered membrane surrounds the lungs. The inner layer is the **pulmonary pleura** which covers the lung. This is then surrounded by the outer **parietal pleura**. The **parietal pleura** are attached to the thoracic walls so that when the thorax expands the lungs expand as well. These two thin epithelial membranes are separated by the pleural space, which contains liquid to allow the pleura to move together without too much resistance but also makes sure the two pleura do not separate. In between the left and right lung and above the **diaphragm** is the heart.

The major airways

Air enters the body via the **upper respiratory tract** that consists of the mouth and nasal cavities, the **pharynx** and the **larynx**. Air will enter the upper respiratory tract via the mouth and/or the nasal cavity. The nasal cavity warms and humidifies the air we breathe and also removes some particulate material. However, during exercise mouth breathing predominates due to the rates of pulmonary ventilation required. The mouth and nasal cavity meet at the **pharynx** and the esophagus splits off at the start of the **larynx**. Here the epiglottis (a flap of cartilage) prevents food from entering the **larynx** and air from entering the esophagus. After the **larynx**, which also contains vocal cords, we enter the **lower respiratory tract** beginning with the **trachea**, which then branches into the left and right **bronchi** supplying the left and right lungs. The **bronchi** branch many times (23) into ever decreasing sizes of bronchioles, which terminate in the alveoli.

Cartilaginous rings support the **trachea** and many of the **bronchi** sub-divisions. These airways also have bands of smooth muscle and contain many **cilia**, little hair-like protuberances, and mucosal glands that secrete fluid to help trap particles. The action of the **cilia** then moves the fluid and particles back towards the upper respiratory tract. **Bronchioles** lack cartilage support and can therefore collapse under pressure. Innervated smooth muscle and elastic fibers as well as secretory glands surround **bronchioles**. Beyond the bronchioles are the **alveolar ducts** and the **alveoli**. The major airways and alveoli are represented in *Fig. 1.*

The alveoli and pulmonary capillaries

Alveoli occur either singularly, as an offshoot from an **alveolar duct,** or in larger numbers in alveolar sacs that form the terminal point of the respiratory airway. There are about 300–500 million alveoli in the lungs and this provides an immense surface area for gas exchange with the pulmonary capillaries. This surface area is often compared to the size of a tennis court. A dense network of pulmonary capillaries surrounds each alveolus. The thin walls of both the alveoli and the capillary mean that gases do not need to diffuse very far in

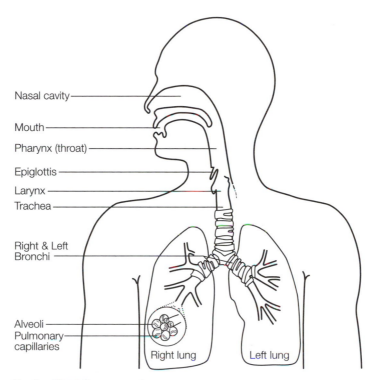

Nasal cavity

Mouth

Pharynx (throat)

Epiglottis

Larynx

Trachea

Right & Left
Bronchi

Alveoli
Pulmonary
capillaries

Right lung   Left lung

Fig. 1.   The pulmonary system.

either direction. The lumen of a pulmonary capillary is very small, allowing only one red blood cell to pass through at a time. This also helps diffusion of gases. The pulmonary capillaries contain blood from the pulmonary artery that is high in carbon dioxide content and low in oxygen content. This is the reverse of the air contained in the alveoli. Diffusion of these gases then occurs down concentration gradients that see oxygen enter the pulmonary capillaries and carbon dioxide enter the **alveoli**.

Muscles associated with inspiration and expiration

A number of muscles form an integral part of the pulmonary system by providing part of the dynamic component required for inspiration and expiration and in some cases providing support and constraints for the **thorax**. The most important muscle involved in pulmonary respiration is the **diaphragm,** which provides the inferior border of the **thorax** and at rest bows slightly upwards. During inspiration the **diaphragm** contracts and this lowers the inferior border of the **thorax** and elevates the ribcage. This lengthening and broadening of the **thorax** increases its volume. The consequence of an increase in lung volume is a decrease in air pressure and therefore air moves into the lungs from the environment to equalize air pressure and the lungs are filled. This process can be helped to a small extent by the **external intercostal** muscles that raise the anterior surface of the ribs. In heavy exercise, with limited time for inspiration, other muscles are also recruited such as the sternomastoid and scalene muscles.

Expiration at rest is a passive process. This involves the recoil of the lung's elastic tissues and relaxation of the **diaphragm,** which will decrease the lung volume. This leads to an increase in air pressure and air will then move back out of the respiratory tract into the environment. During heavy exercise, when expiratory time is decreased, specific muscles can be recruited to help speed expiratory flow. The **internal intercostals** can provide some movement of the ribs and the abdominal muscles can contract to increase abdominal pressure on the underside of the **diaphragm**, which serves to move this muscle further into the **thorax** thus helping to expel air.

The energy cost of the muscle activity associated with respiration is not considered to be large, or limiting, for exercise, even up to maximal aerobic capacity. The **diaphragm** and other muscles associated with respiration are primarily oxidative in nature and thus can support prolonged activity with few signs of fatigue.

Ventilation and lung volumes

The integration of muscular activity and intrinsic (elastic) properties of the lung can combine to increase or decrease total <u>ventilation</u>. **Ventilation** is the total amount of air moved in and out of the lungs over a given period of time. This can be broken down into the breathing or **ventilatory frequency** (the number of times we breathe in a given time period) multiplied by the <u>depth or volume</u> in each breath, referred to as the **tidal volume**. At rest we may go through the inspiration–expiration cycle around 10–15 times per minute. During these cycles the expiratory time exceeds the inspiratory time. When heavy exercise demands a respiratory rate of 50–60 breaths min^{-1} the expiratory period is significantly reduced. The reduction in expiratory time means that the active (muscular) and passive (elastic) properties are both required to meet ventilation demands.

The **tidal volume** is only one component of the volume of air in the lungs. Like respiratory frequency, **tidal volume** can change in different circumstances including exercise. **Tidal volume** at rest may be around 500 ml per breath in a normal healthy adult; however this may change considerably with exercise (*Fig. 2*). **Total lung volume** is the maximum amount of air that the lung can hold. This is normally limited by the size of the individual and thus the size of the **thorax**. Of course at rest our **tidal volume** does not represent our **total lung volume**. **Total lung volume** can be further broken down into the **residual volume** and the **forced vital capacity**. The **residual volume** is the amount of air that cannot be expired no matter how forceful the expiration (not all structures collapse with forced expiration) so some air is left in the lungs at end-expiration. For an average male this may represent around 1000–1500 ml of air

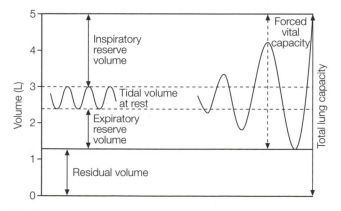

Fig. 2.   Lung volumes.

and for the average female 800–1200 ml. The **forced vital capacity** represents the maximal amount of air that can be ventilated in a single inspiratory–expiratory cycle. This is therefore the maximal **tidal volume** and can reach 3500–4500 ml. The reason that **tidal volume** can increase to such levels is because resting **tidal volume** can expand at the expense of both **inspiratory reserve volume** and **expiratory reserve volume**. When ventilating at rest we do not approach maximal lung emptying or filling. Thus, to cope with demands placed on us during exercise we can breathe more deeply, a process that requires greater lung inflation and deflation.

Other dynamic properties of lung function include those that are time-dependent. With high respiratory rates, inspiratory and expiratory times are significantly reduced. As previously mentioned at these rates the body must be able to breathe in and out quickly and thus the combination of passive and active components of ventilation is important. Dynamic properties can be assessed by looking at the **forced expiratory volume** during the first second of a forced maximal expiration ($FEV_{1.0}$). This volume can also be reported as a percentage of **forced vital capacity** ($FEV_{1.0}/FVC$) and can be used in the diagnosis of obstructive and restrictive lung disorders (e.g. asthma).

D2 PULMONARY FUNCTION AND CONTROL

Key Notes

Pulmonary function

Overall pulmonary function is a complex combination of processes. It includes movement of air in and out of the lungs, ventilation–perfusion matching and the transport of gases in the blood.

Air and gas partial pressures

Gases that are contained in the air we breathe exert a pressure based on the percent volume of that gas and the total gas pressure. This partial pressure will help to drive the process of diffusion of gases between the alveoli and pulmonary capillary blood.

Diffusion

Diffusion is the process by which gases are exchanged between the alveoli and the pulmonary capillary blood. This occurs rapidly due to minimal structural barriers and partial pressure gradients.

Gas transport

Gas transport in the blood can be accomplished in a number of ways. The primary mode of transport for oxygen is via binding to hemoglobin in red blood cells. The binding of oxygen and hemoglobin at the lungs and the release of oxygen from hemoglobin at the tissues are influenced by a range of factors. Carbon dioxide is predominately transported as bicarbonate ions.

Control of breathing

The control of the rate and depth of breathing is primarily neurally mediated. Conscious control is limited and overridden by reflex control from respiratory centers in the brain.

Related topics

Pulmonary structure and volumes (D1)	The neural system (F1)
Cardiovascular structure (E1)	Exercise at altitude (I4)
	Exercise underwater (I5)

Pulmonary function

Pulmonary function reflects more than just the simple mechanical inflation and deflation of the lungs and the generation of **ventilation**. To achieve appropriate pulmonary function there must be optimum diffusion of gases across the alveoli–capillary membranes. To understand these processes we need to evaluate gas pressures, blood perfusion of the lung capillary network, transport of gases in the bloodstream and finally the control mechanisms for pulmonary function as a whole.

Air and gas partial pressures

The air we breathe is made up of a range of different gases and small particulates. The primary gases of interest are **nitrogen, oxygen** and **carbon dioxide**. **Nitrogen** accounts for c. 79.0% of air by volume, **oxygen** c. 20.9% by volume and **carbon dioxide** c. 0.03% by volume. Each of these gases exerts a pressure dependent upon this concentration and the total pressure of the mixed gases.

Given that sea level atmospheric pressure is c. 760 mmHg, then **oxygen** exerts a partial pressure (PO_2) of c. 159 mmHg and **carbon dioxide** exerts a partial pressure (PCO_2) of c. 0.2 mmHg. Gases within the human body are dissolved in fluids and because gas solubility and temperature are relatively constant in blood, the key factor in the exchange of gases between the alveoli and pulmonary capillaries is the gradient of the partial pressures of the gas in the alveoli (P_AO_2) and the pulmonary capillary blood (P_cO_2).

Diffusion

By the time air reaches the alveoli the PO_2 has dropped to c. 105 mmHg due to the mixing of gases in the lung. This is, however, still greater than the PO_2 of blood entering the pulmonary capillaries which is c. 45 mmHg. Diffusion of gases between the alveoli and pulmonary capillaries represents movement down a significant concentration gradient through the single cell wall (epithelium) of both the alveoli and the pulmonary capillary. Gas diffusion is, therefore, quite rapid. Red blood cells move through capillaries fairly slowly (c. 750 ms) and by the time the blood leaves the capillary and enters a venule the PO_2 has risen to match alveoli PO_2, 105 mmHg. This blood is now termed 'oxygenated'.

Due to differences in PCO_2 the diffusion of **carbon dioxide** occurs in the opposite direction to the diffusion of **oxygen** at the **alveoli**. In this instance the air in the **alveoli** has a PCO_2 of c. 40 mmHg. The pulmonary blood entering the capillaries has a PCO_2 of c. 45 mmHg, primarily due to the metabolic production of **carbon dioxide**. Although a smaller gradient than oxygen, the greater PCO_2 in the capillary blood drives **carbon dioxide** diffusion back into the alveoli. **Carbon dioxide** in fact diffuses more quickly than **oxygen** and thus by the time the capillary blood enters a pulmonary venule the PCO_2 has dropped to c. 40 mmHg.

Total diffusion capacity not only depends on the partial pressure gradients in the alveoli and pulmonary capillary blood, but also requires a matching of the **ventilation** with adequate blood **perfusion** (supply) to all areas of the lung. At rest the upper regions of the lung are not well perfused with blood, primarily due to the effects of gravity. Thus diffusion may be limited in these regions due to inadequate blood flow. When we begin to exercise one-way diffusion capacity increases simply by increases in blood flow and thus **perfusion** of the upper areas of the lung.

Gas transport

Diffusion will maximize the partial pressure, and therefore content, of **oxygen** in the pulmonary vein. This **oxygen** must now be transported to all areas of the body. **Oxygen** is transported in two ways; either dissolved in plasma or bound to **hemoglobin** in red blood cells (see Section E). The amount dissolved in plasma is very small (2% of total) and thus makes a small contribution to the supply of **oxygen** to the body's cells. The amount attached to **hemoglobin** (98% of total) amounts to c. 20 ml **oxygen** per 100 ml of blood. Each 100 ml of blood contains c. 16 g of **hemoglobin** (slightly lower levels in women) and each **hemoglobin** molecule can bind 1.34 ml of **oxygen**.

Four **oxygen** molecules can bind to each **hemoglobin** complex to form **oxy-hemoglobin**. The ability of **hemoglobin** to bind **oxygen** is dependent on a range of factors but is usually represented by an **oxy-hemoglobin** dissociation curve (*Fig. 1*). When the partial pressure of oxygen (PO_2) is high, as it is in the alveoli and then pulmonary capillaries, the **hemoglobin** becomes almost fully saturated with **oxygen**. As the PO_2 decreases, as it does in tissue such as skeletal

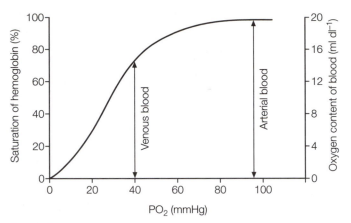

Fig. 1.   The oxy-hemoglobin dissociation curve.

muscle, then **oxygen** is released from **hemoglobin** and is taken up by the local tissue. Factors that affect the **oxy-hemoglobin** curve include temperature and **pH,** which reflects the balance of acidity and alkalinity. A lower **pH** causes greater unloading of **oxygen** and this occurs at the exercising muscles whereas the **pH** is normally high in the lung promoting **oxyhemoglobin** binding. Within the red blood cell itself the production of 2,3-diphosphoglycerate also reduces the affinity for oxygen binding with hemoglobin and promotes oxygen unloading at tissues during exercise.

The removal of **carbon dioxide** from the body, primarily via the lungs, requires that it be transported from the tissues to the lungs. This can occur in three ways; either dissolved in plasma, bound to **hemoglobin** as **carbamino-hemoglobin**, or as a **bicarbonate ion** (HCO_3^-). The HCO_3^- originates from the dissociation of carbonic acid:

$$(H_2O + CO_2 \leftrightarrows H_2CO_3 \leftrightarrows H^+ + HCO_3^-)$$

The **carbon dioxide** dissolved in plasma is only a small proportion of the total amount of carbon dioxide in the body and is easily exchanged at the lungs where there is a lower partial pressure in the alveoli. **Carbamino-hemoglobin** is so-called because **carbon dioxide** can bind to **hemoglobin** via amino acids in the globin part of the molecule. **Carbon dioxide** binding is affected by the PCO_2 such that binding is promoted in tissues with high partial pressures and unloading is promoted in the lungs. The primary mechanism of transport is as part of a HCO_3^- ion and this accounts for c. 70% of the total **carbon dioxide** transported. Carbonic acid (H_2CO_3) is bound together very weakly and thus quickly dissociates to a hydrogen ion and a HCO_3^- ion. The hydrogen ion can bind with the **hemoglobin**, which shifts the **oxy-hemoglobin** dissociation curve to the right promoting oxygen unloading. **Hemoglobin** in this way is also acting as a buffer for the hydrogen ion and preventing a decrease in **pH** that would occur if hydrogen ions accumulated. In the lungs the hydrogen ion and HCO_3^- reform carbonic acid that then dissociates to water and **carbon dioxide** which diffuses into the alveoli and is expelled in the breath.

Gas exchange at the tissues is primarily related to the metabolic state of that tissue. Active tissue will release heat and hydrogen ions that will precipitate the unloading of **oxygen** from hemoglobin, which can then be taken up by the active tissue. Once in the cells oxygen is transported to **mitochondria** to be used in the full oxidative metabolism of carbohydrates, fats or proteins. From 20 ml of **oxygen** per 100 ml of blood the body removes about 5 ml per 100 ml of blood at rest and this is termed the **arterio–venous oxygen difference**. At maximal exercise the body may remove c. 17 ml of **oxygen** per 100 ml of blood. This increased **oxygen** extraction combines with greater blood flow to promote massive increases in **oxygen** consumption when healthy adults approach maximal exercise intensities. **Carbon dioxide** produced in the tissue simply diffuses out into the bloodstream down a partial pressure gradient and is then transported to the lungs to be exhaled.

Control of breathing

Control of breathing is a complex process about which much has still to be learned. In essence what the body continues to do in the control of breathing and ventilation is to maintain the homeostasis of PO_2 and PCO_2. Control processes are primarily neurally mediated (see Section F).

We can exert a small degree of conscious control over breathing (have a try now by either increasing or decreasing your rate or depth of breathing for a just few seconds). Most control, however, reflects involuntary activity of the inspiratory and expiratory respiratory centers in the **medulla oblongata** and **pons** areas of the brain. These are powerful control centers and can override conscious control, which prevents us from holding our breath too long! The areas govern both the breathing frequency and the tidal volume of each breath by neural links to the respiratory muscles. When altering breathing depth and rate the control centers are responding to central (motor cortex) and peripheral sensory information (*Fig. 2*). There is clearly a key role for motor cortex control of ventilation especially in situations such as exercise where rapid alterations in **ventilation** are required. Other important information comes from **chemoreceptors**. For example, **chemoreceptors** in the aortic arch or the carotid bodies respond to changes in the PO_2 in the blood as well as HCO_3^- concentration and PCO_2. In this way we see a clear match between total **ventilation** and the requirements to maintain PO_2 during lower levels of incremental exercise (i.e. we breathe enough to get the right amount of **oxygen** delivered to the working muscle cells). Interestingly, and this point will be explored further in later sections (D3), at higher exercise intensities ventilation seems to be coupled to the need to expire **carbon dioxide** to regulate its partial pressure. Thus at

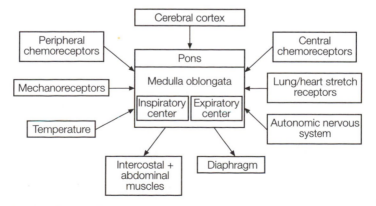

Fig. 2. Control of ventilation.

higher levels of exercise there is a relative hyperventilation with respect to **oxygen** delivery needs.

Other neural control comes from stretch receptors located within the lung and heart and further input may come from mechanoreceptors in joints, tendons and active muscles that assess movement or metabolic status.

D3 PULMONARY RESPONSES TO EXERCISE

Key Notes

Overview	Whatever the exercise mode, duration and intensity, there will be a co-ordinated ventilatory response. This response will likely precede exercise initiation. The response will persist throughout the exercise and into the recovery period.
Rest and low exercise intensities	Ventilation demonstrates a two-phase response to exercise that includes a primary rapid adjustment and a secondary slower adjustment in total ventilation. The primary phase is likely driven by central neural commands and the secondary phase likely represents an on-going adjustment to blood chemical status.
Moderate and high exercise intensities	Beyond a certain exercise intensity, generally referred to as the anaerobic threshold, ventilation rises in line with carbon dioxide removal and over and above the rate of oxygen consumption. The mechanisms behind this are controversial but likely reflect anaerobic metabolism in the muscle.
Maximal exercise and maximal voluntary ventilation	Data suggest that ventilation levels at maximal exercise intensity do not reach maximal voluntary ventilation rates. This suggests that in normal healthy humans and in normal environmental conditions ventilatory factors do not limit exercise performance.
Aging, gender and disease	Age, gender and disease can all affect the ventilatory response to acute exercise. However, despite declining ventilatory function with age and smaller lung volumes and hemoglobin levels in women it is unlikely that ventilation is a limiting factor in exercise performance. Such a limit will only likely be reached in normal people under severe environmental stress or with underlying pulmonary disease.

Related topics	Energy sources and exercise (B1)	Cardiovascular responses to
	Energy for various exercise	exercise (E3)
	intensities (B5)	The neural system (F1)
	Pulmonary structure and volumes	Exercise at altitude (I4)
	(D1)	Exercise fitness and health (K1)
	Pulmonary function and control	Screening and exercise prescription
	(D2)	(L1)
	Cardiovascular structure (E1)	

Overview   Exercise will stimulate **ventilation**. The reason, as alluded to on many occasions, is that an increase in ventilation is a prerequisite for an increase in **oxygen** consumption. As the first step in the '**oxygen cascade,**' **ventilation** will increase alongside cardiovascular adaptations, to be outlined in Section E, to deliver substantially more **oxygen** than at rest to the working muscles.

Rest and low exercise intensities

As **ventilation** is neurally mediated there can be a small anticipatory effect as the body prepares, however this is not too large as excessive hyperventilation is not good preparation for exercise. As the body starts exercising at low intensities (30–50% of maximum aerobic capacity) the response of the respiratory system is to increase total **ventilation** to match the metabolic needs, or **oxygen** delivery to the muscle. This is achieved primarily by increasing the depth of breathing via neural stimulation of the inspiratory muscles, predominantly the **diaphragm**. The ventilatory response to the onset of exercise is also described as 'two-phase' with an initial rapid increase in **ventilation** brought about by motor cortex activity and feedback from active musculature and joints routed to the respiratory center. The slower second phase of ventilatory adaptation to exercise reflects the sensor activity that occurs in response to temperature and partial pressures of gases in the blood. Other sources of neural input may come from the heart.

At low exercise intensities the link between ventilatory volume and **oxygen** uptake is quite closely coupled and as such there is no excessive **hyperventilation**. This relationship is often described by the ventilatory equivalent for oxygen (ventilatory volume divided by oxygen consumption) which generally stays fairly stable over low–moderate exercise intensities. The ventilatory equivalent for **carbon dioxide** (ventilatory volume divided by volume of **carbon dioxide** produced) also stays relatively stable at low exercise intensities due to the matching between **ventilation** and aerobic metabolic activity.

Moderate and high exercise intensities

At moderate to high exercise intensities the volume of **ventilation** will continue to increase partially to meet the demands of the muscle for **oxygen** delivery. However, it is clear from *Fig. 1* that ventilatory volume begins to rise out of line with increases in **oxygen** uptake and thus we see a relative **hyperventilation** (i.e. there is an increase in the ventilatory equivalent for **oxygen**). This relative **hyperventilation** is primarily achieved through increases in respiratory rate and volume and thus depends not only upon active inspiration but also active expiration as well. The ventilatory equivalent for carbon dioxide initially stays fairly stable as the ventilatory equivalent for oxygen rises. This period is called isocapnic buffering. After this the ventilatory equivalent for carbon dioxide will also rise.

What can be clearly seen from *Fig. 1* is a breakpoint in ventilation that closely matches that of **carbon dioxide** production rather than **oxygen** consumption. What is likely happening, although this is controversial, is that higher levels of PCO_2 are controlling and thus elevating ventilatory volume rather than PO_2. Debate continues as to the exact nature of the control of exercise-related **hyperventilation**. Many suggest it is a response to excess **carbon dioxide** production as a result of **lactic acid** metabolism in active muscle tissue (see Section B5). **Lactic acid** in skeletal muscle very quickly disassociates to a lactate molecule and a hydrogen ion. The hydrogen ion can be buffered in the muscle and blood by HCO_3^-, which results in the 'non-aerobic' release of **carbon dioxide**. The ensuing change in ventilation, which occurs at about 55–75% of maximal aerobic capacity in most people, was described in the 1960s as the **ventilatory threshold** or **ventilatory breakpoint** (also referred to as the anaerobic threshold). Much excitement greeted this initial work as it was deemed that

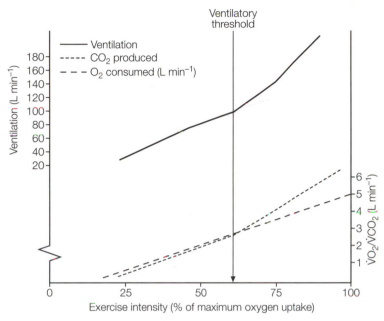

Fig. 1. Ventilatory responses to graded exercise.

the **ventilatory breakpoint** was coincident with and caused by the **lactate breakpoint** or **lactate threshold** in response to graded exercise. Thus it was surmised that ventilatory assessment could be used to non-invasively determine the **lactate threshold**.

Initially the **ventilatory breakpoint** was simply selected from a plot of **ventilation** or volume of **carbon dioxide** produced against exercise intensity or **oxygen** consumption. Other methods have been used that have included the simplistic assessment of when the respiratory exchange ratio (volume of **carbon dioxide** produced divided by volume of **oxygen** consumed) rose above one. Ventilatory equivalents have also been used with the ventilatory equivalent for oxygen demonstrating a sharp upturn before any change in the ventilatory equivalent for **carbon dioxide** (see *Fig. 2*).

Maximal exercise and maximal voluntary ventilation

At maximal exercise in a normal healthy adult male or female the volume of **ventilation** may reach c. 125 L min^{-1} (slightly lower in women). This represents an enormous increase from a resting value of c. 6 L min^{-1}. This is possible due to a very high breathing frequency, up to c. 50 breaths per minute and an increased tidal volume of c. 2.5 L breath^{-1}.

The question that is often posed is whether this level of **ventilation** actually represents the limit of pulmonary function and therefore could lung function be a potential limitation to the transfer of **oxygen** and thus maximal aerobic capacity. It is clear in most individuals that at maximal aerobic exercise there is still some capacity left in the amount of air ventilated in and out of the lungs. This is easily demonstrated in a maximal voluntary ventilation test. This test, performed at rest, normally requires the subject to consciously maximally ventilate for 15 seconds and then the volume measured is extrapolated to a minute (it is not wise to hyperventilate like this for any longer than 15 seconds). It is

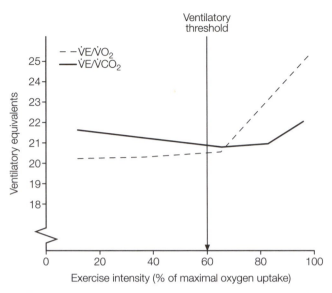

Fig. 2.   Ventilatory equivalents for oxygen and carbon dioxide during graded exercise.

clear that this conscious effort exceeds the values on minute ventilation seen at maximal aerobic capacity. Due to this fact it has often been assumed that ventilatory function cannot be a limiting factor to maximal aerobic capacity. Indeed as well as the maximal voluntary ventilation data it is clear that PO_2 in pulmonary blood returning to the left heart is nearly always 105 mmHg and the percentage saturation of **hemoglobin** with **oxygen** rarely drops below c. 98%, even at maximal exercise. The weight of evidence therefore does not suggest a limiting role for lung function and **oxygen** diffusion in normal healthy adults. The picture may be different in highly trained endurance athletes or those with pulmonary disease.

Aging, gender and disease

Pulmonary function is primarily related to the size of the **thorax**, which will determine the size of the lungs and thus the alveolar cross-sectional area. To this extent lung volumes and lung function have been closely related to height and body surface area. Other factors that will impinge on ventilatory function relate to disease and the changes in body tissues associated with aging.

The aging process results in many changes to the structure and function of the body. Whilst it is difficult to specifically differentiate the effects of aging from a concomitant rise in disease rates and declining levels of activity it is possible to speculate on the impact of aging upon pulmonary function. One aspect of aging is a decline in stature and this would likely impact upon static lung dimensions that underpin dynamic lung function. Associated with this are changes in the elasticity of tissues, due to increasing accumulation of connective tissue, which may result in changes in the expiratory function of the lungs. The overall results of aging-related changes will be blunted **ventilation** at maximal aerobic capacity.

The primary impact of gender on ventilatory function is related to the smaller average body size of females compared to males. However, it is also pertinent to note that women have on average 4 g per 100 ml less **hemoglobin** than men and thus the oxygen-carrying capacity of blood in females is lower. Despite this there is little evidence that respiratory factors or **oxy-hemoglobin** factors limit maximal aerobic capacity in women.

The biggest likely impact upon pulmonary function comes from either an extreme environment (altitude) or disease processes. We will refer to the impact of exercise at high altitude (with a reduced PO_2) in Section I4. There are many diseases that can affect pulmonary function including chronic illnesses such as bronchitis and emphysema. One specific problem of interest to exercising children and adults is the impact of **asthma**, as its prevalence is seemingly on the increase. **Asthma** is a disease that causes a narrowing of the **bronchioles**, which leads to a restriction in the ability to ventilate (*Fig. 3*). There are many triggers for an **asthma** attack and exercise is one such factor. During an **asthma** attack breathing becomes much more difficult due to the narrowing of the airways and this can lead to painful breathing (**dyspnea**). Various prophylactic drug options are available and for athletes with **asthma** careful management should not severely restrict activity. Performing activity in warm and relatively humid conditions helps prevent exercise-related **asthma** attacks, as does a thorough warm-up procedure.

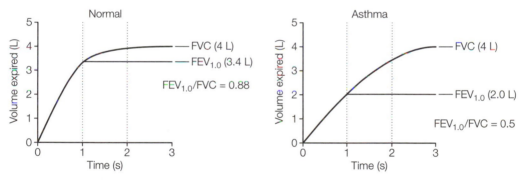

Fig. 3.   Forced expiratory volume in one second in a normal subject and a subject with asthma.

D4 PULMONARY RESPONSES TO TRAINING

Key Notes

Responses at rest and low exercise intensities
Only small changes in pulmonary function are seen at rest and low exercise intensities after training. This is likely due to the fact that there is the same oxygen demand for a given absolute exercise intensity, irrespective of training state.

Responses at moderate and high exercise intensities
Pulmonary function responses to graded exercise are different above and below the ventilatory breakpoint. Below the pre-training breakpoint only small changes are seen post-training, however, above this level (due to a shift in the ventilatory breakpoint) ventilation is lower at absolute exercise intensities.

Responses at maximal exercise
At maximal exercise ventilatory volume can be increased drastically after training, often to over 200 L min^{-1}. This is accomplished due to a combination of a greater tidal volume and respiratory rate. In some individuals this may push pulmonary function to a level where it might limit maximal aerobic capacity.

Aging, gender and disease
The trainability of the pulmonary system is not affected by aging and gender. There is evidence that exercise training in patients with pulmonary disease can improve respiratory symptoms and improve overall exercise capacity.

Related topics

Energy sources and exercise (B1)
Energy for various exercise
   intensities (B5)
Pulmonary structure and volumes
   (D1)
Pulmonary function and control
   (D2)
Pulmonary responses to exercise
   (D3)

Cardiovascular structure (E1)
Cardiovascular responses to
   training (E4)
The neural system (F1)
Training principles (H1)
Exercise at altitude (I4)
Exercise fitness and health (K1)
Guidelines for exercise prescription
   (L2)

Responses at rest and low exercise intensities
In individuals who undergo systematic endurance exercise training there are relatively few noticeable pulmonary adaptations at rest. Static lung volumes such as total lung volume change little with training. This should not be surprising, as we do not grow in stature to any great extent with training. **Forced vital capacity** may increase slightly and this is likely a response of the respiratory muscles to training rather than a change in lung size per se. If **forced vital capacity** increases slightly the likely consequence is a small decrease in **residual volume**.

At rest or in low–moderate exercise intensities there is some evidence of a small reduction in respiratory rate. As a consequence ventilation at rest or at the same absolute exercise intensity may decline very slightly, suggestive of greater ventilatory efficiency. Such small changes are not surprising when one considers that resting and submaximal exercise intensities require the same **oxygen** delivery pre- and post-training. If the energy cost and thus **oxygen** demand of any given exercise is the same after training then the body must ventilate to a similar level to meet that demand.

Responses at moderate and high exercise intensities

Tidal volume, respiratory rate and **ventilation** are fairly similar pre- and post-training across exercise intensities below the anaerobic threshold. Again, any small adaptations may be a small reduction in **ventilation** due to a reduced respiratory rate and a slightly greater tidal volume. However, it is clear from research data that exercise training can produce a right shift in the **anaerobic threshold** and its constituent components the **lactate breakpoint** and the **ventilatory breakpoint** (*Fig. 1*).

For any exercise intensity that was above the **ventilatory breakpoint** before training and is then below the ventilatory breakpoint post-training, the ventilatory volume will be reduced. This reduction in ventilation is primarily due to a decrease in respiratory rate as the body does not yet perceive the drive to expel the excess **carbon dioxide** being liberated as part of anaerobic metabolic activity.

Responses to maximal exercise

Despite minimal changes in ventilation pre- and post-training, at rest and during low-intensity exercise there is evidence of substantial changes in **tidal**

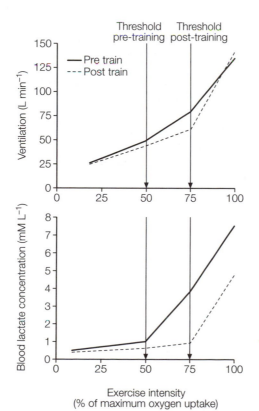

Fig. 1. Lactate and ventilatory breakpoints pre- and post-training.

volume, respiratory rate and ventilatory volume at maximal aerobic exercise. Maximal respiratory rate and maximal **tidal volume** increase post-training for a profound effect upon maximal ventilatory volume. Indeed some of the bigger endurance athletes, notably rowers and cyclists, have been reported to produce a ventilation volume of >200 L min^{-1} at maximal exercise intensities. Combined with cardiovascular changes post-training (an increase in maximal **cardiac output** and **arterio–venous oxygen difference)** changes in maximal ventilation underpin the increased performance capacity and maximal aerobic power in elite endurance athletes.

In normal healthy individuals pulmonary function is unlikely to be a limit to maximal aerobic capacity. This is also the case for most trained individuals. However, in highly trained endurance athletes there is some evidence of a relative **oxy-hemoglobin** desaturation at maximal exercise. This means that not all available hemoglobin is combining with oxygen in the pulmonary capillaries. The likely causes of this are speculative but may include poor **ventilation–perfusion** matching and/or a problem with rapid red blood cell transit time in the pulmonary capillaries. If such time is limited because of high flow rates associated with maximal exercise then there may not be enough diffusion time for alveolar and capillary blood PO_2 to equilibrate. This would lead to a decrease in **oxygen** saturation and content in a given volume of blood. Interestingly, this issue would not hamper carbon dioxide exchange because of its faster diffusion speed.

Aging, gender and disease

There is little evidence to suggest that the qualitative changes that occur in ventilation with training differ between young and old and between males and females. In absolute terms adaptations in older adults and in females will be smaller than in young male adults but this is again a likely consequence of their initial bigger body size.

Exercise training in specific disease states may be very beneficial for improving performance at low and moderate exercise as well as completing activities of everyday living. In these instances this positive outcome is likely a consequence of a shift upwards in the exercise intensity that would lead to discomfort and **dyspnea**. This is possibly related to more efficient pulmonary function and a drop in respiratory rate at the same absolute exercise intensities as well as a general training effect.

E1 CARDIOVASCULAR STRUCTURE

Key Notes

The circulatory system	The circulatory system is vital to human function at rest as well as being integral to the ability to adjust to the demands of exercise and training. The primary function of the circulation is to pump blood to virtually all tissues of the body in order to supply oxygen and other substrates. Blood also has a function in removing waste products from cells.
The heart	The heart is a muscular organ that is the connection between the pulmonary and systemic circulatory systems. The heart produces contractile force that moves blood around the body.
Blood vessels	Blood vessels include the large arteries and veins, smaller arterioles and venules as well as the very small exchange capillaries surrounding cells such as in skeletal muscle. The size and structure of different blood vessels reflects their specific roles.
Blood	Blood is the complex fluid within the circulatory system. Blood's primary role is to transport oxygen and other substrates around the body. Blood has many different constituents. The simplest subdivision of blood is when it is split into plasma content and cellular content.

Related topics

The circulatory system

The primary role of the **circulatory system** is to deliver, via the bloodstream, **oxygen** and other substrates to metabolizing tissues. Other important roles include the removal of waste products from tissues, the (re)distribution of **heat** to maintain **thermal balance** and the movement of other substances (e.g. **hormones**) around the body to target sites and organs. To perform these roles the circulatory system is very responsive to the demands placed upon it but it is also quite fragile and vulnerable to disuse and disease.

The circulatory system is a 'closed-loop' system that consists of a central pump (the **heart**), two circulatory branches, the **pulmonary circulation** and the **systemic circulation** (see *Fig. 1*) and a transport medium (the **blood**). The major difference between the pulmonary and systemic circulatory systems is that the pulmonary system is a lower-pressure system that supplies blood to the **alveoli** in the **lungs** for the exchange of gases and water, whereas the systemic circulation is a higher-pressure system that supplies every other tissue and organ in the body. Both circulatory systems consist of 'delivery' blood vessels (**arteries and arterioles**), 'exchange' vessels that allow the passage of nutrients between

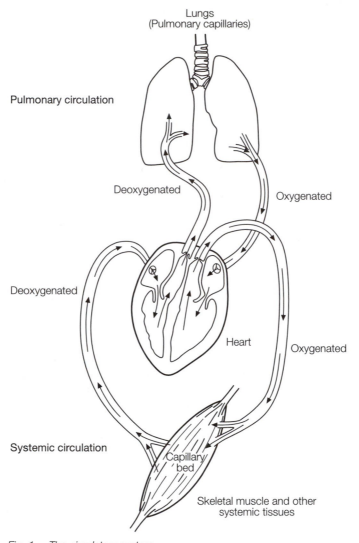

Fig. 1.   The circulatory system.

intra- and extravascular spaces (**capillaries**) and 'collecting' vessels bringing blood back to the heart (**veins** and **venules**).

The heart

The heart is situated in the middle of the chest cavity and is protected by the sternum and ribs. It is roughly the size of a clenched fist and comprises mainly cardiac muscle cells. The heart receives blood from major veins and then contracts to force blood into the major arteries. This is achieved through a controlled and synchronized contraction and relaxation process. *Fig. 2* demonstrates that the heart comprises four chambers; these include two **ventricles** (left and right) and two **atria** (left and right). The ventricles are the largest chambers and the forceful contraction of blood from these chambers results in blood flow into the systemic and pulmonary circulation via the aorta and pulmonary artery, respectively. When blood returns to the heart from the systemic

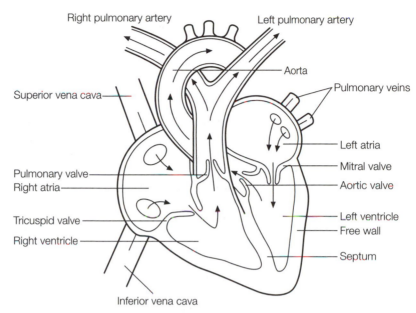

Fig. 2. *The heart.*

circulation (via the **venae cavae**) and the pulmonary circulation (via the **pulmonary vein**) it is first emptied into the right and left atria, which are smaller collecting chambers that sit directly above the ventricles. As well as receiving blood the thin-walled atria also pump blood into the ventricles.

Surrounding the chambers are the cardiac walls made up primarily of cardiac muscle cells. The orientation of these cells allows contraction in three dimensions (longitudinal axis, short axis and rotational). Cardiac muscle cells are similar to skeletal muscle cells (see Section C1) in that they contain contractile proteins and appear striated when viewed microscopically. However, cardiac muscle cells are: (i) more highly interconnected through intercalated discs, or gap junctions, that allow more rapid propagation of electrical activity between cells; (ii) contain some special cells that display autorhythmicity, which means depolarization and thus contraction can occur without external electrical signals; (iii) shorter and thinner; (iv) have a single nucleus in the center of the cell; and (v) contain more mitochondria than skeletal muscle cells.

The other major structures in the heart are the valves. The aortic and pulmonary valves help regulate flow between the ventricles and the major arteries. Atrio-ventricular valves include the mitral valve that separates the left atrium from the left ventricle and the tricuspid valve that separates the right atrium from the right ventricle. All the valves are thin membranes that passively open and close in response to changes in pressure. The other major role of valves is to prevent regurgitant or 'back' flow of blood.

Blood vessels

The key differences, structurally, between the five major types of blood vessels (arteries, arterioles, capillaries, venules and veins) reflect their role and anatomical location. Large arteries receive blood at high pressure and to cope with this they have a relatively thick wall. This wall includes the **endothelium,** which is the inner wall of the vessel. A further layer of **elastin** allows large arteries to stretch when receiving blood and then recoil which helps propagate flow through the arterial system. The next layer is **smooth muscle** that can contract and relax to help regulate the size of the vessel lumen and thus alter pressure and flow. The final layer is **collagen**, a structural protein that helps maintain the integrity of the vessel (see *Fig. 3*).

Arterioles have relatively more smooth muscle and this reflects their key roles of **vasodilation** (increased cross-sectional area of the vessel lumen) and **vasoconstriction** (decreased cross-sectional area of the vessel lumen) that help regulate blood flow to various capillary beds. Arteries and arterioles have smaller total cross-sectional area than the capillaries but are under the greatest pressure with higher blood flow velocity reflecting their 'delivery' role. The **capillaries** are the smallest vessels and they consist of a single layer of endothelial cells. This allows exchange of substances between the capillaries and cells. When total vessel cross-sectional area increases in the capillaries, because of the sheer number of them, both blood flow velocity and blood pressure drop and this enhances greater transport of substances into and out of the capillaries. **Venules** and **veins** have smaller amounts of elastin than arteries. This means they are more distensible and can change shape to accept large volumes of blood with very little changes in blood pressure. Most of the blood volume in a human is contained within **venules** and **veins**. Blood pooling after a period of exercise will occur primarily in the veins of the lower limbs. **Veins** and **venules** contain smooth muscle so **vasoconstriction** can occur to help blood flow back to the heart.

Blood

There is a total **blood volume** of c. 4–6 L in the human body, dependent primarily on body size and fitness level. Blood serves many functions most of which relate to its cellular constituents and the contents of plasma. Plasma is a straw-colored liquid that is clearly visible if blood is centrifuged or 'spun' after being collected. Plasma is primarily made up of water but does contain coagulation proteins (e.g. fibrinogen), hormones and many other substances. If left the

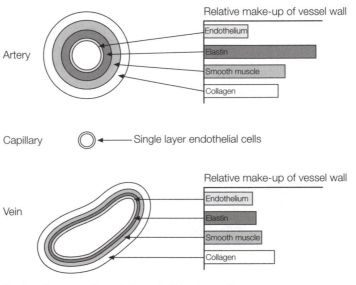

Fig. 3. Cross-sections of the major blood vessels.

proteins in plasma will coagulate, as fibrinogen is converted to fibrin, and the remaining liquid is called **serum**.

There are many types of cell in blood of which the most numerous is the red blood cell (**erythrocyte**). There are smaller numbers of a range of different white blood cells (**leukocytes)** and **platelets** (thrombocytes).

Red blood cells are biconcave disks whose lifespan is c. 120 days. They are produced in the bone marrow and then migrate to the circulation. The process of red cell formation is called **erythropoiesis**, which is controlled by the hormone **erythropoietin**. One of the most important constituents of a red blood cell is the hemoglobin complex that gives the cell, and blood, its red color. The primary role of the red blood cell is the transport of oxygen and carbon dioxide bound to the heme complex, containing iron, in hemoglobin. From both a health and sporting context changes in red blood cell number and volume (as assessed by the **hematocrit**) must be carefully monitored. A reduction in **hematocrit** is referred to as hemodilution. If **hematocrit** levels drop too low it is called anemia and may represent disease and/or poor nutrition. Some athletes will attempt to increase their **hematocrit** and thus **oxygen** delivery capacity by natural (altitude training) or by unnatural means (**erythropoietin** injection or **blood doping**). In both these instances the **hematocrit** may rise substantially beyond the normal level of 40–45% of total blood volume.

Leukocytes constitute a much smaller percent of blood volume (c. 1%) but their importance is not diminished by their size or relative volume. Five types of leukocytes have been identified and these are (in descending order of cell numbers); neutrophils, lymphocytes, monocytes, eosinophils and basophils. The primary role of the leukocytes is in the defense of the body against foreign organisms. They achieve this in two ways. Firstly neutrophils literally destroy bacteria by surrounding them and chemically breaking them down (phagocytosis). Second, lymphocytes play a key role in the immune system response to invading cells. The immune system is extremely complex but other substances such as antibodies play an important role.

Platelets are numerous but small and represent a very small percent of total blood volume. Platelets interact with substances released from vessel walls and circulating blood coagulation factors to prevent blood loss after a vessel is injured.

E2 CARDIOVASCULAR FUNCTION AND CONTROL

Key Notes

The electrocardiogram (ECG)

The function of the heart relies upon the electrical signals that pass between the cardiac muscle cells. It is only after the cardiac muscle cells are depolarized by a 'wave' of electrical activity that the muscle cells contract and blood is actually pumped. The ECG represents a global picture of electrical activity in the heart.

The cardiac cycle

The integration of electrical activity, muscle contraction, and change in volume, pressure and flow are integral components of a single cardiac cycle.

Cardiac output and stroke volume

The flow that occurs from both ventricles as a result of electrical and mechanical activity is best represented by two variables, cardiac output and stroke volume. Cardiac output is the total volume of blood flow from a ventricle per unit of time (normally a minute) and stroke volume is the amount or volume of flow from a ventricle produced per heart beat.

Peripheral blood flow

Peripheral blood flow refers to all blood flow that occurs outside of the heart. Blood flow in the periphery is dependent upon the ejection force, the reflective pulse wave of the elastic arteries as well as vasodilatation and vasoconstriction that occur primarily in the arterioles and venules.

Venous return

The primary variable of interest in terms of blood flow returning to the heart from the tissues is the venous return. This will determine preload, or the amount of filling of the ventricles, and is influenced by a range of factors.

Related topics

Cardiovascular structure (E1)	Cardiovascular responses to training (E4)
Cardiovascular responses to exercise (E3)	The neural system (F1) The endocrine system (F2)

The electrocardiogram (ECG)

Blood flow requires a highly regulated and integrated contraction of cardiac muscle cells that is governed by electrical signals. Each cardiac cell contracts only after the propagation of an action potential that depolarizes the cell. This precedes muscle contraction and the electrical repolarization of the cell prior to the next action potential.

The heart is unique in that some specific cells can initiate the process of depolarization independently of external influences. In this way the heart is said to be autorhythmic. The initial depolarization occurs in the sino-atrial node which is a small area located in the right wall of the right atrium (see *Fig. 1*). This is a pacemaker cell that has an unstable resting membrane potential that will spontaneously depolarize at regular intervals. As the sino-atrial node produces an

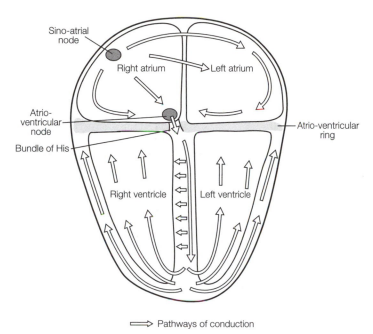

Pathways of conduction

Fig. 1.   The pathway of electrical depolarization in the heart.

action potential so a wave of depolarization is spread across the atria. This process is rapid and both atria contract together. An insulating fibrous ring separates the atria from the ventricles so electrical conduction between the upper and lower chambers only occurs through one pathway, the atrio-ventricular node. This is a relatively slow-conducting part of the system, which allows full ventricular filling prior to ventricular depolarization. The electrical signal then passes through the bundle of His and the bundle branches in the septal wall before spreading out through the Purkinje fibers into the ventricles. This is very rapid allowing both ventricles to contract together.

The integration of all cardiac cell action potentials can be assessed at the surface of the body via electrocardiography (ECG; see *Fig. 2*). The ECG is split into various component parts and periods. The most recognizable nomenclature is PQRST. The P component or P-wave represents the simultaneous depolarization of the left and right atria. This a small wave due to the relatively small mass of the atria. A small delay (P–R interval) represents the electrical signal's transmission through the atrio-ventricular node. The QRS complex or wave is normally a thin but large spike of electrical activity that represents simultaneous depolarization of the left and right ventricles. The larger spike represents a greater muscle mass and the short duration reflects the speed of transmission of electrical activity through the Purkinje fibers. The QRS complex masks the repolarization, or recovery of the atrial cells. A short S–T segment represents the refractory period of ventricular cells, which is followed by the T-wave representing the ventricular repolarization. A period of baseline electrical membrane potential precedes the next P-wave. Apart from the determination of heart rate, the number of beats per minute of the entire myocardium, the ECG can provide direct and indirect evidence of health and disease within the heart. Because of this the ECG is an invaluable clinical tool.

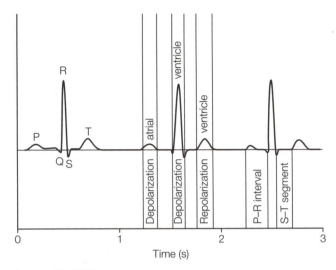

Fig. 2.   The ECG.

Electrical conduction pathways are also sensitive to external control (see Section F). The primary target for external influence is the sino-atrial node and various factors alter the rate of spontaneous depolarization that will either speed up or slow down the heart rate. The cardiovascular centers in the brain are linked to the sino-atrial node via the sympathetic and parasympathetic pathways of the autonomic nervous system. Sympathetic nervous activity increases the heart rate and parasympathetic activity decreases heart rate. At rest both pathways are active but the parasympathetic pathway is dominant (**vagal tone**). Parasympathetic neural activity is transmitted to the sino-atrial node via the neurotransmitter acetylcholine. The sympathetic neurotransmitter, noradrenaline (norepinephrine), and the hormone adrenaline (epinephrine) released from the adrenal gland under sympathetic neural control can both speed depolarization and thus heart rate.

The cardiac cycle

The ECG is the representation of the electrical activity within the heart. Depolarization is followed by contraction, which is the basis for pressure development within the heart. Pressure changes result in valve opening (and closing) and precipitates blood flow.

The cardiac cycle in *Fig. 3* integrates the time sequence of electrical activity, changes in pressure, flow and volume of blood in the left side of the heart. Beginning at the onset of the P-wave the heart is still in the period called **diastole** or relaxation of the ventricles. At this point the pressure in the aorta is at its lowest and the pressure in the left atria and ventricle is close to zero. The atrio-ventricular valve is open due to the small pressure gradient between the atria and ventricle and the ventricular volume is quite high due to early passive diastolic filling. Consequent to the P-wave is atrial **systole** (contraction), which increases the atrial pressure and results in active blood flow from the atria to the ventricle that adds to ventricular volume. At rest 2/3 of ventricular filling is early/passive whilst only 1/3 of filling is as a consequence of atrial systole. After a short electrical delay the onset of the QRS complex occurs which is the rapid depolarization, synchronously, across the ventricles (although the septum

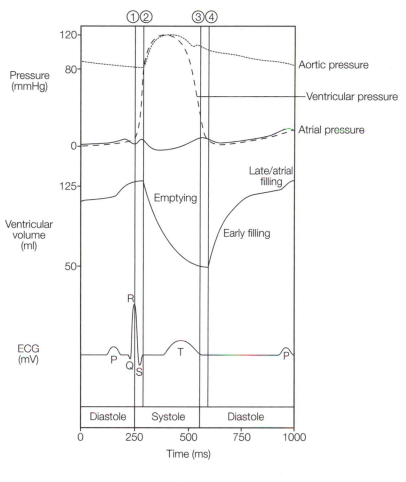

Fig. 3.   The cardiac cycle.

depolarizes slightly before the rest of the ventricles). After the electrical depo-larization comes the mechanical event of ventricular systole. As soon as muscle contraction occurs ventricular pressure starts to rise which results in the closure of the mitral valve. The aortic valve remains shut for a short period as ventricu-lar pressure rises to meet aortic pressure. This period when both the mitral and the aortic valves are shut is called the **iso-volumetric contraction** period or pre-ejection period. There is a small and transient rise in left atrial pressure as the mitral valve and atrio-ventricular ring is pushed slightly up compressing the atrium. As soon as the rapidly climbing ventricular pressure exceeds aortic pressure the aortic valve is opened and blood flow occurs out of the left ventri-cle. Ventricular pressure continues to rise as the contraction continues. As blood is ejected aortic pressure rises in parallel with ventricular pressure and ventric-ular volume rapidly decreases. After ventricular contraction has finished and repolarization has begun the impulse behind blood flow into the aorta is removed and ventricular pressure begins a rapid drop. Shortly after falling below the aortic pressure the aortic valve shuts to prevent blood flow back into the ventricle. This valve closure signals the end of systole and the beginning of diastole. However, the ventricular pressure must drop some more before it falls below atrial pressure when the mitral valve will open. Again a period of aortic and mitral valve closure occurs called **iso-volumetric relaxation**. As soon as ventricular pressure drops below atrial pressure the mitral valve opens and the passive early filling blood flow occurs into the ventricle and so begins the refilling process that is completed with atrial systole at the end of diastole.

Cardiac output and stroke volume

Every time the ventricles contract they produce blood flow. This is called the **stroke volume (SV)**. The number of times the heart beats per minute is called the **heart rate (HR)**. The combination of heart rate and stroke volume produces **cardiac output ($\dot{Q}$)**, which is the key functional variable of the heart. **Stroke volume**, the consequence of a single cardiac cycle, is controlled by a range of factors that includes ventricular filling or **preload**, resistance to flow, or **afterload**, and the intrinsic contractile properties and processes of the ventricle, or **contractility** (often referred to as inotropic state). The concept of preload and its impact on stroke volume was described in detail by Frank and Starling when they stated that if the ventricle is pre-stretched by a greater filling then the contraction force or tension generated during systole is increased and both **stroke volume** and **ejection fraction** (the ratio of stroke volume to left ventricular volume at end-diastole) increase. Thus a link exists between preload and contractility. An increase in preload and thus stroke volume can be seen when you lie down from standing, during exercise, or with an increase in blood volume. Afterload is the pressure against which the ventricle has to work to produce flow. For the left ventricle this can be assumed to be equivalent to pressure in the aorta. If aortic pressure were to increase then the ventricle would spend longer contracting and generating pressure in the ventricle before the aortic valve would open and hence stroke volume would decrease. The body's normal response to an increased blood pressure (e.g. going from lying to standing) is to increase heart rate to maintain cardiac output.

Contractility may be partially determined by the length of the cardiac muscle cell as already described but contractility can also increase under the influence of autonomic control. The sympathetic nervous system not only innervates the sino-atrial node but also directly innervates the ventricles and thus can increase contractility. A rise in sympathetic neural tone (see Section F), such as with exercise, will directly lead to a greater contractility and greater ejection fraction.

Peripheral blood flow

The factors that control peripheral blood flow are complex and not fully understood. One of the most important sites of blood flow control is the arteriole. As the major arteries branch into many arterioles the resistance to flow is great. Control of flow is achieved by the contraction or relaxation of smooth muscle in the vascular walls of the arteriole, which leads to **vasodilatation** and **vasoconstriction**. This can markedly alter flow as well as pressures in the arteries and capillaries.

The process of **vasodilatation** and **vasoconstriction** is under both intrinsic and extrinsic control. Intrinsic mechanisms include autoregulation that may be related to changes in wall tension and/or metabolite release. Substances such as ATP, ADP, adenosine, lactate, pyruvate as well as potassium ions have been implicated as vasodilators that help couple blood flow with increased metabolic activity. Additionally, nitric oxide is a powerful vasodilator that is released from the endothelial cells of blood vessels in response to increase vessel wall stress. Extrinsic mechanisms include neural and hormonal factors (see Section F). In many tissues an increase in sympathetic activity will cause a general vasoconstriction. The neurotransmitter noradrenaline (norepinephrine) has a potent vasoconstrictor effect (via α1 adrenoreceptors) in tissues such as skeletal muscle, the skin, the gut and the kidneys. Noradrenaline (norepinephrine) can cause a weak vasodilatation (β2 adrenoreceptors) but vasoconstrictive effects swamp this. Within skeletal muscle some sympathetic nerves innervate arterioles via the neurotransmitter acetylcholine and this results in a vasodilatation. Whilst not often active it may be important in exercise when skeletal muscle demands a significant proportion of total cardiac output. Vasodilatation in most tissues is simply brought about by a reduction in sympathetic tone. Additionally, adrenaline (epinephrine) may cause vasodilatation via β2 adrenoreceptors in skeletal muscle. Other vasoactive hormones include angiotensin II, antidiuretic hormone and atrial natriuretic peptide.

Flow within the capillaries is regulated 'up-stream' by the activity of arteriolar smooth muscle and precapillary sphincters. The precapillary sphincters are thought to be controlled by metabolite levels. Capillaries have no ability to individually **vasoconstrict** or **vasodilate** and thus total blood flow through a capillary bed is related to the total number of capillaries that are either open or closed.

Venous return
Venous return is the sum of all blood flowing back to the right atrium from all tissues of the body (apart from the alveoli). The venous system attempts to maintain **venous return** to the right side of the heart at all times. This is because venous return and cardiac output are inextricably linked due to the fact that circulation is a closed-loop system. You cannot pump out what is not delivered to the heart. The primary factor influencing venous return is the driving pressure from the arterial side of the circulation. Often mean circulatory filling pressure, and its gradient with right atrial pressure represent this. Like other blood vessels, **veins** and **venules** are under the influence of the sympathetic nervous system (see Section F) that will lead to vasoconstriction. Circulating levels of adrenaline (epinephrine) and angiotensin II will have the same effect. In addition venous valves prevent backward flow in the venous drainage of the lower limbs and rhythmic contraction and relaxation of skeletal muscles, called the **skeletal muscle pump**, provides a driving force for venous flow. The heart itself provides some suction when right atrial pressure fluctuates below zero at specific points in the cardiac cycle. A respiratory pump also promotes venous return as the cycle of inspiration and expiration results in periods of negative pressure within the thoracic cavity. This results in a moderate increase in venous flow during inspiration.

E3 CARDIOVASCULAR RESPONSES TO EXERCISE

Key Notes

General impact of exercise	Any acute exercise bout will elicit an integrated organ response to meet the metabolic demands, primarily in the skeletal muscle. In this regard the most important goal is to increase oxygen delivery to the muscle to match the ATP demand of the exercise.
Cardiac output	Cardiac output is the product of heart rate and stroke volume. Cardiac output is the primary functional variable of the heart. This represents the central pump function and thus the blood flow available to the whole body. Cardiac output will increase with the onset of dynamic exercise.
Heart rate	Heart rate is the number of times the heart beats in a minute. Changes in heart rate with exercise such as tachycardia (a quickening of heart rate) are called chronotropic responses.
Stroke volume	Stroke volume is the volume of blood ejected per beat. Stroke volume is under the control of preload, afterload and contractility in the normal healthy heart. In dynamic exercise stroke volume will increase from rest.
Peripheral blood flow	Peripheral blood flow relates to blood flow in all vessels outside of the heart. Peripheral blood flow adjusts to exercise in order to increase oxygen delivery to the active skeletal muscle.
Exercise intensity, duration and mode	Various types of exercise, either in the form of short, high-intensity bursts of activity or prolonged submaximal steady rate exercise, will exert a range of different cardiovascular responses.
The impact of aging and gender	Aging results in a general deterioration in cardiovascular function and structure. The qualitative response to exercise is similar in men and women. Absolute cardiovascular responses to exercise will also reflect the average increase in body size and muscle mass seen in men.

Related topics	Work and power performed on the cycle ergometer and treadmill (A3)	Cardiovascular responses to training (E4)
	Estimation and measurement of energy expenditure (A4)	The neural system (F1)
		The endocrine system (F2)
	Cardiovascular structure (E1)	Thermoregulation (I1)
	Cardiovascular function and control (E2)	Exercise in hot and humid environments (I2)

General impact of exercise

When the exercise physiologist tries to analyze factors important in determining oxygen uptake we tend to refer to the **Fick equation** (see Section A). It is clear from the **Fick equation** that oxygen delivery and consumption requires an integrated response from both the central (cardiac output) and peripheral circulation (arterio–venous oxygen difference). $\dot{V}O_2 = \dot{Q} \times a - \dot{V}O_2$ difference.

Cardiac output

During exercise, cardiac output increases from resting values of approximately 6 L min⁻¹ to reach maximal values of 20–30 L min⁻¹ or perhaps greater depending on body size and training status. During rhythmic exercise across a range of exercise intensities there is a close coupling between **cardiac output** and **oxygen** consumption as can be seen in *Fig. 1a*. We also know that **cardiac output** is the product of **heart rate** and **stroke volume** so the changes in these variables with exercise will determine the **cardiac output** response (see *Fig. 1b and 1c*).

Heart rate

The **heart rate** at rest is c. 70 beats min⁻¹, although this can range from 30–100 beats min⁻¹, dependent upon training status and other factors. **Heart rate** increases in a linear fashion with **oxygen** consumption to the point where

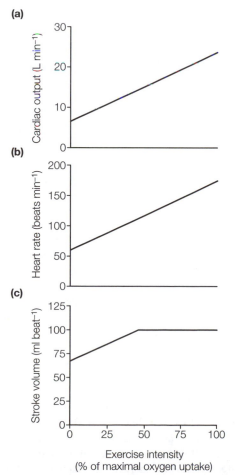

Fig. 1.   The response of cardiac output, heart rate and stroke volume to graded exercise.

oxygen uptake reaches a peak or maximal value. At this point **heart rate** will be maximal as well. We can predict this through a simple and common equation where:

$$\text{maximal heart rate} = 220 - \text{age}$$

The primary mechanisms for an increase in **heart rate** with exercise are related to neural and hormonal control. At the onset of exercise the parasympathetic neural activity is reduced and this alone will result in an increase in **heart rate**. Subsequently, sympathetic neural drive is increased and this will also increase **heart rate**. Adrenaline (epinephrine) will also cause the **heart rate** to rise.

Stroke volume

Stroke volume at rest is c. 80 ml beat^{-1}, although this is influenced by gender, body size, training status and posture among a range of factors. Stroke volume increases with the onset of exercise, however, there is some debate about the kinetics of stroke volume change with increasing levels of exercise intensity. Most research would suggest the stroke volume plateaus at about 40–50% of exercise capacity (c. 100–150 ml beat^{-1}). The reason provided is the decreasing filling time available as heart rate accelerates. Conversely some researchers have suggested that stroke volume actually increases in a linear fashion all the way up to maximal exercise. This has been observed primarily in trained individuals but remains contentious.

The mechanisms underpinning stroke volume changes with exercise are twofold. First, as cardiac output increases so the mean circulatory filling pressure increases and coupled with processes such as the skeletal muscle pump, **venous return** will increase. This will increase filling of the ventricles that will result in an elevated end-diastolic volume and a larger stroke volume. Second, the increased sympathetic neural drive also influences ventricular muscle cells to increase the force of contraction.

Peripheral blood flow

At rest the skeletal muscle receives about 20% of cardiac output. This may rise to up to 80% of cardiac output at maximal exercise and is achieved by a complex combination of vasoconstriction, in relatively inactive or less important tissues such as the gut, the kidney and inactive skeletal muscle, and vasodilatation in the active muscle beds. Blood flow distribution is controlled by neural input and local regulation. Specifically alterations in shear stress in response to increased flow and the change in the biochemical make-up of intra- and extravascular fluid will produce rapid and significant alterations in flow through arterioles, capillaries and in the venous system.

One of the primary variables associated with peripheral blood flow is that of blood pressure. Blood pressure provides the driving force for flow via pressure gradients. Arterial blood pressure is commonly measured and is generally reported as its two constituent components; systolic and diastolic blood pressure. Systolic blood pressure changes in response to exercise from resting values of c. 120 mmHg to values c. 200 mmHg at maximal aerobic exercise intensity. This reflects a greater stroke volume and cardiac output being pumped into an arterial system that has limited capacity to distend and stretch. Diastolic blood pressure, the residual pressure in the arteries during ventricular relaxation, changes little from a resting value of 80 mmHg. Small changes in diastolic pressure reflect the balance of vasoconstriction and vasodilatation, primarily in the arterioles, upon peripheral resistance to blood flow.

Exercise intensity, duration and mode

Exercise intensity can vary from very low intensities (such as walking), through maximal aerobic exercise (associated with maximal oxygen consumption) and up to supramaximal exercise intensities (such as sprinting). The impact of changes in exercise intensity is that there is an inverse relationship with exercise duration until fatigue. For example, sprinting is very high-intensity exercise, however, we can only do this for brief periods of time due to limitations in substrate availability or metabolic waste product build-up. The response of cardiovascular function to short bouts of such activity will likely be different to sustained sub-maximal aerobic exercise because of the limited time for cardiovascular adjustment as well as the high level of contractile force generated by skeletal muscle. Such contractile force can actually increase the external pressure on blood vessels to the point that vessels are constricted and flow is reduced or stopped. This leads to a reduced delivery of blood to the tissues and hence a decline in venous return. Consequent to this is an increase in afterload or blood pressure due to an elevated peripheral resistance. This is best seen in high-intensity weight-lifting where blood pressures may reach over 300 mmHg during both systolic and diastolic phases of the cardiac cycle.

Alternatively, some people participate in ultra-endurance events such as triathlons. These people may exercise sub-maximally for as many as 12 hours and such prolonged activity will influence cardiovascular function. For example the loss of plasma volume through a large sweat loss as well as the necessity to lose heat via vasodilatation of skin blood vessels may both reduce central blood volume and thus mean circulatory filling pressure. A consequence of this would be a reduced ventricular filling and thus a decrease in **stroke volume**. To maintain cardiac output would require an increase in heart rate. The progressive rise in heart rate with prolonged exercise, called **cardiovascular drift** (see *Fig. 2*), is a common phenomenon that has also been recently linked to direct temperature effects in the sino-atrial node and elevated sympathetic activity (see also Section I2).

Another variable, exercise mode, may also alter the cardiovascular responses to exercise. For example the gravitational effects on the cardiovascular system during standing exercise (running) are different to seated exercise (cycling) and horizontal or supine exercise (swimming). Changes in posture will influence blood flow and neural activity. Supine exercise will elicit higher **stroke volumes**, because of enhanced preload, and thus lower **heart rates**.

If a small muscle mass is exercised, such as the arms (compared to the legs), then the cardiovascular system will respond differently. In arm exercise (even if

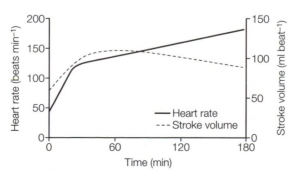

Fig. 2. Cardiovascular drift in prolonged exercise.

rhythmic in nature) the peripheral resistance is increased and this will increase blood pressure because of a smaller capillary and arteriolar bed. These changes will decrease **venous return** and produce a greater increase in **heart rate** than during leg exercise at the same absolute work rate.

The impact of aging and gender

There are many factors that influence the cardiovascular response to an acute exercise bout. The influence of training will be dealt with in Section E4 and factors such as time of day and environmental conditions will be addressed later in the text (Section I). Two other factors that are known to affect the cardiovascular responses are age and gender.

A key component of aging is that maximal heart rate is age-dependent and thus we see a reduction in **heart rate reserve** (maximal heart minus resting heart rate) with aging that will limit maximal cardiac output and performance capacity. The exact mechanism of this change is not known. Other changes include a slow reduction in the number of contractile cardiac cells with age. This affects both systolic and diastolic function in the heart including a decrease in maximal stroke volume that further reduces cardiac output. Peripheral changes with age include a decrease in major artery compliance and this increased arterial stiffness will result in an increase in resting and exercise blood pressures. The chronic effect of elevated blood pressures is a stimulus for an increase in cardiac muscle mass, which then increases myocardial oxygen demand. It is, however, important to note that the true effects of healthy aging are often masked under a myriad of cardiovascular diseases whose prevalence increases with age.

In most cases, men are bigger and have less adipose tissue than women and these factors have a large influence on cardiac size and function. When cardiac size is corrected for lean mass (primarily muscle) there is evidence that gender differences in cardiac size and performance diminish or disappear altogether. Heart rate is not gender dependent and despite some evidence to the contrary the qualitative responses of heart rate, stroke volume and cardiac output to progressive exercise are similar in males and females.

There is some evidence that resting and exercise blood pressures are lower in women, at least up to the age of the menopause when the cardio-protective influence of estrogen is lost. Any further differences in exercise capacity between males and females will likely represent a combination of a lowered hemoglobin concentration and thus oxygen-carrying capacity in women and subtle changes in metabolic characteristics in the exercising muscle.

E4 CARDIOVASCULAR RESPONSES TO TRAINING

Key Notes

Responses at rest	The effect of endurance-based training on the cardiovascular system includes some adaptations that are noticeable at rest. An example of this would be a decrease in resting heart rate.
Responses to submaximal exercise	The adaptations in the cardiovascular system at rest are carried over to sub-maximal exercise. These changes underpin many performance improvements that occur with training. An important change is an increase in stroke volume at any absolute work rate.
Responses to maximal exercise	Trained individuals achieve higher work rates and maximum oxygen uptake. These changes are underpinned by increases in maximal stroke volume, cardiac output and arterio–venous oxygen difference.
Training modality	Changes in the cardiovascular system are dependent on the nature of the training. This is probably most marked when considering the responses to endurance training compared to resistance or weight training.
The impact of aging and gender	The responses to exercise training in older, compared to younger subjects, and females, compared to males, has been the subject of continuing debate and research. Most believe the qualitative responses to training are similar in nearly all human sub-populations.

Related topics

Cardiovascular structure (E1)
Cardiovascular function and control (E2)
Cardiovascular responses to exercise (E3)

The neural system (F1)
The endocrine system (F2)
Training principles (H1)
Training for aerobic power (H2)

Responses at rest

One of the most common responses to endurance training is a reduction in resting heart rate. This can often manifest itself in a heart rate below 60 beats min^{-1}, which is referred to as a **resting sinus bradycardia**. Interestingly, cardiac output at rest is maintained at pre-training levels and thus an increase in resting stroke volume offsets the decrease in heart rate.

The primary stimulus for a decrease in heart rate is an increase in para-sympathetic tone to the sino-atrial node. There is, however, evidence that sympathetic tone is also decreased and the intrinsic heart rate may also be reduced. The change in stroke volume may reflect alterations in heart rate that will increase diastolic filling time, an increase in blood volume and a larger left ventricular cavity. There is considerable evidence from cross-sectional athlete–control studies as well as longitudinal chronic exercise designs that left ventricular cavity size is increased with training. Despite this information, the question as to whether the decrease in heart rate is a response to the increase in stroke volume or vice-versa is not fully known.

There is limited evidence of any reduction in systolic and/or diastolic blood pressure post-training unless the subjects were hypertensive prior to the start of training. These changes may be related to a reduction in arterial stiffness.

Responses to submaximal exercise

At the same absolute exercise intensity cardiac output will be similar pre- and post-training with a reduced heart rate matched by an increased stroke volume. When exercise intensity is adjusted to be relative to the maximal oxygen consumption cardiac output and stroke volume data are still raised after exercise training compared to pre-training data. In this scenario heart rate remains the same pre–post training (see *Fig. 1*).

It is clear that if the subjects exercised at the same **relative** exercise intensity pre- and post-training the individual would be working harder or faster after training. For example, if a subject worked at an exercise intensity that elicited 50% of heart rate maximum pre- and post-training then the higher stroke

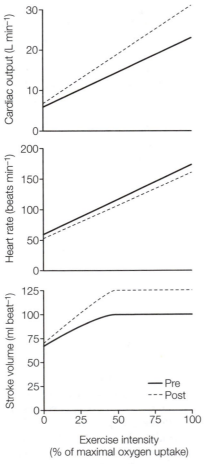

Fig. 1.   Cardiac output, stroke volume and heart rate responses to graded exercise pre- and post-endurance exercise training.

volume after training would result in a higher cardiac output meeting a higher metabolic demand at the muscle.

Responses to maximal exercise

Maximal heart rate is not affected by training to any great extent although some data exist to suggest slightly lower maximum heart rate post-training possibly due to a blunted sympathetic drive. At maximal exercise the increased cardiovascular and performance capacity is primarily related to an increased stroke volume and thus maximal cardiac output. The mechanisms have been previously mentioned but the importance of enhanced diastolic filling parameters may be even more important at the high heart rates associated with maximal exercise. Blood pressure data post-training again are relatively similar to pre-training values.

There is also a component of peripheral blood flow adjustment to training that contributes to the enhanced exercise capacity post-training. Data suggest a modest increase in the ability to extract oxygen as assessed by **arterio–venous oxygen difference**. This parameter, c. 5 ml oxygen per 100 ml of blood at rest can increase to maximal exercise values of c. 15 ml oxygen per 100 ml of blood. After training an increase up to 17 ml oxygen per 100 ml of blood has been reported. This is likely to be due to changes in the ability to preferentially re-route blood flow to active muscle tissues, a greater capillarization of active skeletal muscle beds and the enhanced oxygen extraction capability of the trained muscle cells with greater numbers of mitochondria and oxidative enzymes.

Training modality

Whilst endurance training has been the focus of this chapter due to its common use in health and sporting contexts, other forms of training may have different effects on the cardiovascular system. With weight-training the key component of performance is the rapid development of skeletal muscle force. Thus performance is not influenced or limited by the adjustments or capacity of the cardiovascular system. This is in contrast to the endurance-training model where a constant and high cardiac output is vital to the delivery of metabolic substrates to support muscle activity. The classic changes in heart rate, stroke volume and cardiac output with endurance training are not seen with resistance training unless the training has some significant aerobic component as seen in circuit training. An interesting pattern of cardiac muscle mass adaptation to training has been observed in the scientific literature. In this instance the larger cavity dimension seen in endurance-trained individuals is replaced by a thicker wall surrounding a normal cavity with resistance training (see *Fig. 2*). Whilst slightly controversial, a hemodynamic theory for this training-specific pattern has been proposed. In endurance training the ventricular cavity dilation is due to replication of sarcomeres in series in response to the end-diastolic wall stress due to increased filling with exercise. Conversely, the wall thickening with weight training likely represents an adaptation or addition of sarcomeres in parallel to normalize the end-systolic wall stress due to elevated blood pressure responses to exercise.

The impact of aging and gender

Older adults do not lose the ability to respond to training stimuli. Recent data have suggested subjects as old as 80–90 years can benefit from endurance training with improvements in performance capacity. The likely magnitude of change with training may decline in older subjects. Of some interest are recent data that suggest that whilst the functional capacity improvements witnessed in trained older males and females may be similar the cardiovascular changes

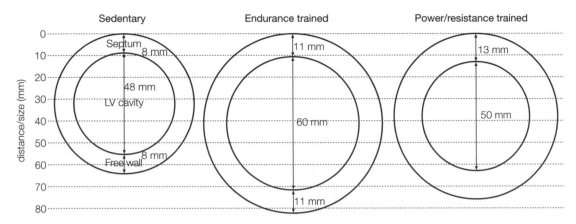

Fig. 2.   Left ventricular size adjustments to different types of training.

underpinning these adaptations may be different. In men contractile and structural adaptation have been reported whereas women seem to rely predominantly upon peripheral adaptations. The reasons for this disparity are not clear at this time.

Most data suggest that younger males and females demonstrate the same qualitative patterns of cardiovascular adjustment to training such as reductions in heart rate mirrored by increases in stroke volume. A reduced quantitative increase in ventricular cavity size and stroke volume in women is seen as a consequence of a smaller initial cavity size and stroke volume.

F1 THE NEURAL SYSTEM

Key Notes

Overview and homeostasis

The human body monitors and controls its own internal environment in an attempt to maintain homeostasis even in the face of the stress of exercise. The way in which the body maintains homeostasis at rest and during exercise is via the interaction of the neural and endocrine systems. Together the neuro-endocrine system allows humans to respond to a broad range of exercise and sporting endeavors.

Role of the neural system

The neural system is a rapid response control mechanism that detects relevant information, integrates multiple sources of information and initiates appropriate action to restore homeostasis.

Structure of the neural system

The neural system has a central component as well as a peripheral component that can be split into the somatic and autonomic nervous systems. Further the autonomic nervous system has both sympathetic and parasympathetic branches.

Neural system reception, transmission and activation

A range of receptors detect information about the internal and external environment. These generate neural activity that is directed to the central nervous system. Transmission of neural activity requires a range of neurotransmitters. A range of receptors at the target cells modulates activation from a neural signal.

Neural activity during exercise

Both the somatic and the autonomic nervous systems are activated during exercise. The somatic system regulates muscle contraction. An increase in sympathetic neural activity, allied to a decrease in parasympathetic activity, promotes changes in many tissues that support the physiological response to exercise.

Neural adaptations to training

Although difficult to directly assess there is indirect evidence that training alters the acute response of the somatic and autonomic nervous systems to exercise.

Related topics

Energy sources and exercise (B1)
Control of energy sources (B4)
Energy for various exercise
    intensities (B5)
Responses to training (B6)
Muscle structure (C1)
Motor-neural control of contraction
    and relaxation (C2)

Pulmonary adaptations to exercise
    (D)
Cardiovascular adaptations to
    exercise (E)
Exercise and environmental stress
    (I)

Overview and homeostasis

Control of human physiological and biochemical processes has already been developed in various sections of the book. For example in Section B4 we were introduced to the control of the activation of energy sources that are used to fuel the contracting skeletal muscle. Likewise in Section C2 we reviewed the neural control of skeletal muscle contraction and relaxation. Further in Sections D2 and E2 we were introduced to the control and regulation of the pulmonary and cardiovascular systems respectively. However, it was felt appropriate that we should revisit the topic of control and monitoring of the response to exercise. This will provide a deeper insight into physiological control during exercise as well as underpinning later sections.

Control of human physiological and biochemical processes is essential. What the body cannot allow is a lack of control of internal processes. For example, internal body temperature cannot be allowed to fluctuate wildly in exercise and increase to the point where tissues are damaged or destroyed. The body prevents this and thus maintains homeostasis. Homeostasis is therefore often referred to as a state of fluctuating control or balance in the body's internal environment. When exercise is initiated homeostasis in physiological and biochemical processes is disrupted. Muscle contraction raises muscle tempera-ture and metabolism results in alterations in factors such as pH that are a threat to the internal environment and the safe functioning of the human body. In an effort to maintain homeostasis during exercise the human responds with changes in pulmonary and cardiovascular function to redistribute heat around the body as well as to increase oxygen delivery to muscles that will help prevent an over emphasis on anaerobic metabolism as well as prevent the build up of carbon dioxide.

Despite evidence of local control, or autoregulation, in some physiological and biochemical processes, control of the human response to exercise is mainly contained within the actions of neural networks and the release of hormones (the endocrine system). Not surprisingly, these systems overlap and overall control of the human exercise response is often referred to as being under neuro-endocrine control. The two systems work co-operatively to maintain homeostasis and share many features. An example is that both rely on chemical compounds to communicate with target cells (neurotransmitters and hormones). The primary difference between the two systems is related to the speed of response. The neural system's response is more rapid and is important for both initiating exercise and in the physiological adaptation to the onset of exercise in an attempt to maintain homeostasis (e.g. rapid increase in pulmonary and cardiovascular function). The delivery of hormones to target cells takes more time and as such the endocrine system is more important in the maintenance of homeostasis over longer periods of exercise and into recovery.

Role of the neural system

The neural system is a fast-acting control mechanism for a huge number of human physiological and biochemical processes. Like any monitoring system it has three general components: (i) a monitoring or sensory system that detects relevant information around the body that relays messages (neural action potentials) to the central nervous system; (ii) an integrative control center that assimilates all relevant information; and (iii) a response system that sees neural activity sent to specific target tissues to initiate a response that would seek to maintain homeostasis (see neural control of motor function in Section B2).

Structure of the neural system

Much of the detail of the specific structure of the neural system has been presented in Section B2. That section explained the nature of the neuron, the nerve impulse, the synapse and the neuromuscular junction. This material was presented to explain the entire process of activation of skeletal muscle contraction as a fundamental building block of exercise. Here we seek to build on this information and provide a more general picture of the entire neural network.

Overall the neural system has two major components; the central nervous system (CNS) and the peripheral nervous system (PNS) (see *Fig. 1* and also Section B4). The CNS acts as the integrator of information received from the PNS. The first role of the PNS is detection of information, via multiple receptors, and transmission of this information, via afferent nerves, to the CNS. The CNS acts as an integrator and interpreter of information received from the PNS before initiating a response through the PNS that transmits neural activity, via efferent nerves, to target tissues.

The PNS can be split into the somatic system and the autonomic nervous system (ANS). The somatic system connects the CNS to the skeletal musculature and thus is vital in voluntarily initiating muscle contraction and movement. The ANS, which is under involuntary control, connects the CNS to cardiac muscle, smooth muscle and endocrine glands and thus has a significant role in the metabolic and physiological adaptation to exercise. The ANS can be split into two branches, the sympathetic nervous system and the parasympathetic nervous system. These two branches of the ANS have diametrically opposite effects on their target tissues and cells. Generally we consider the sympathetic nervous system as the one that prepares us for flight, fight and fright. It up-regulates activity in the cardiovascular and pulmonary systems and promotes an increase in cellular metabolism. Conversely, the parasympathetic system works to slow down metabolism and return the body to its resting state.

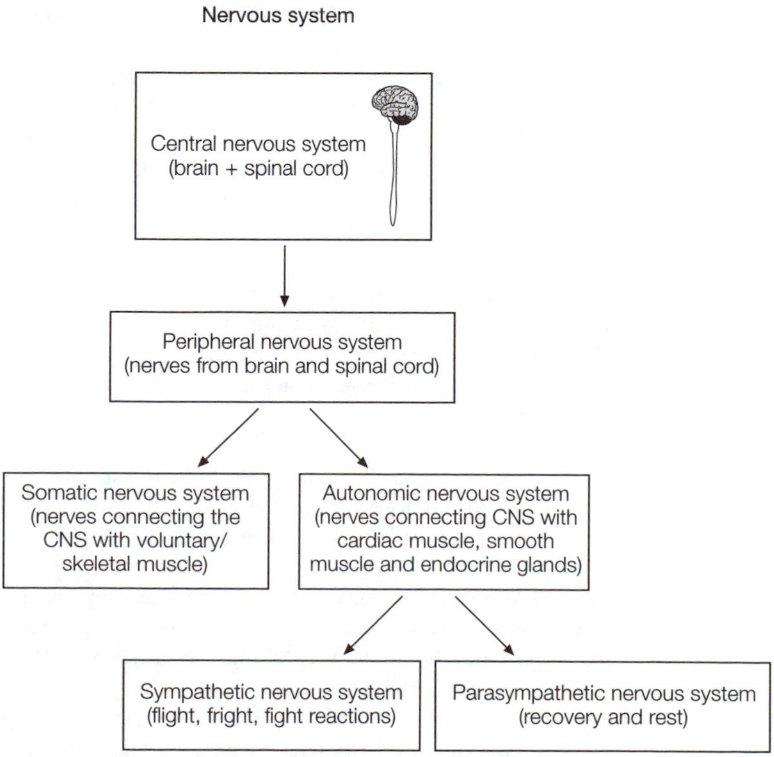

Nervous system

Fig. 1.   Organization of the nervous system.

Neural system reception, transmission and activation

The neural system relies on a number of different types of receptors to provide information to the CNS via afferent neurons. These include a number of sensory receptors throughout the body (that detect sound, sight, smell, touch, taste, pressure, pain and temperature), mechanoreceptors and proprioceptors in the muscle, tendon and joints of the musculo-skeletal system (that detect pressure, stretch, contraction, orientation) and chemoreceptors, for example, in the cardio-vascular system at the aortic arch and carotid bodies (that detect chemical constituents such as PO_2, PCO_2 and pH). Mechanoreceptors and proprioceptors can produce a reflex action in skeletal muscles, via the spinal cord section of the CNS that does not require conscious control from the brain. All other receptor activity travels back to the brain for central processing.

Neural activity is transmitted by nerve cells (neurons) via action potentials. Information passes from neuron to neuron or from nerve cell to target cell across synapses via chemical messengers called neurotransmitters. A range of different neurotransmitters is used to transmit neural activity across synapses in different systems around the body. Acetylcholine is the neurotransmitter used in the somatic nervous system and in the neuron-to-neuron transmission of action potentials in both branches of the ANS. Acetylcholine is also the neuro-transmitter between the neuron and the target cells in the parasympathetic branch of the ANS. Neurons that release acetylcholine are called cholinergic. The neurotransmitter between neurons and target cells in the sympathetic branch of the ANS is normally noradrenaline (norepinephrine). Neurones that release noradrenaline (norepinephrine) are called adrenergic.

Different types of receptors in target cells can produce different responses to neural activation. There are two types of receptors for cholinergic neurones: nicotinic and muscarinic. Similarly there are two types of adrenergic receptors: alpha (α) and beta (β), which are further differentiated as $\alpha1$ and $\alpha2$ as well as $\beta1$ and $\beta2$. Many target cells have more than one type of receptor and this reflects that many human physiological and biochemical processes can be both up- and down-regulated. For example, cardiac cells are innervated by parasym-pathetic neurons (muscarinic) and sympathetic neurons (adrenergic, $\beta1$). The parasympathetic neurons will be activated when heart rate and contractility are decreasing whereas sympathetic neurons are activated to increase heart rate and contractility.

The binding of a neurotransmitter with a receptor at the target cell will induce a number of changes in the target cell or tissue, such as the alteration of cellular membrane potential (e.g. cardiac cells), altering enzyme and thus meta-bolic activity (e.g. adipose tissue) and directly causing muscle contraction (e.g. skeletal muscle).

Neural activity during exercise

During exercise the somatic nervous system has a fundamental role in produc-ing muscle contractions that are the basis of movement and thus exercise. This aspect of neural activity and muscle contraction is covered in Section C2.

The response of the ANS during exercise is much more complex and involves many different target cells. In essence the primary role is taken by the sympa-thetic nervous system although the withdrawal of parasympathetic activity is also important (the analogy is taking the brake off at the same time as pushing down on the accelerator to get a car moving).

Sympathetic nerves innervate various aspects of the pulmonary and cardio-vascular system. Key responses include an increase in bronchiole dilation that will increase airflow (decrease resistance to flow), an increase in heart rate and myocardial contractility that will raise cardiac output and vasoconstriction of blood vessels in non-active muscles that will promote blood flow to working muscles as well as serve to maintain total peripheral resistance and blood pressure. The consequence of all these changes is an increase in oxygen transport to the working muscles as well as increased substrate delivery and metabolite removal from the muscles.

Sympathetic neural activity also promotes sweating and skin blood flow so that exercise will not adversely affect homeostasis through an inappropriate build up of heat in the muscles and core of the body (see Section I).

Sympathetic stimulation of the adrenal gland will promote the release of adrenaline (epinephrine) and noradrenaline (norepinephrine) into the bloodstream. These bloodborne catecholamines have the same effect on the pulmonary and circulatory systems as the sympathetic nervous system. Adrenaline (epinephrine) and noradrenaline (norepinephrine) also promote an increase in fatty acid release from adipose tissue and glucose release from the liver. In this respect the combined sympathetic response to exercise is preparing for exercise and recovery by making metabolic substrates available to the muscles.

The effect of exercise intensity on neural response is relatively straightforward. The greater the exercise intensity the greater the sympathetic activity. There is, however, some evidence of a threshold response in adrenaline (epinephrine) and noradrenaline (norepinephrine) response to graded exercise that has been linked to lactate accumulation in the bloodstream (see *Fig. 2*).

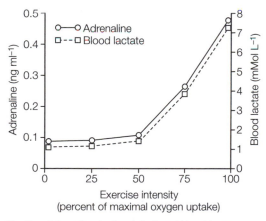

Fig. 2.   Adrenaline (epinephrine) and blood lactate response to graded exercise.

Neural adaptations to training

Neural activity is not particularly easy to measure and so we rely on the assessment of the response of the target cells to acute and chronic exercise. There is no doubt that neural activity is significantly altered by acute exercise but any effect that training has on the neural response to exercise is less clear. There is some evidence of training changes in the somatic nervous system that include alterations towards a more efficient muscle fiber or motor unit recruitment. Adaptations in the ANS are tricky to assess but changes in target tissue responses such as resting heart rate, due to greater parasympathetic tone (see *Fig. 3*) and lowered exercise heart rates at the same external work rate suggest training influences on neural activation.

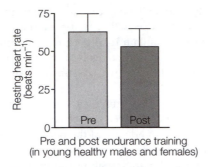

Pre and post endurance training
(in young healthy males and females)

Fig. 3.   Resting heart rates before and after prolonged endurance training.

F2 THE ENDOCRINE SYSTEM

Key Notes

Role of the endocrine system	The endocrine system is a slower-acting control system that interacts with and supports the role of the neural system in maintaining homeostasis in the face of an exercise challenge.
Structure of the endocrine system	Hormones are chemical substances released from a cell or gland that are transported by fluids within the body to a target cell where a physiological effect will occur. Thus the endocrine system is composed of the tissues that produce and release the hormones, the hormones themselves, the transport medium and target cells for the hormones.
Activation and action of the endocrine system	The release of hormones can be promoted by neural, hormonal or other bloodborne stimuli. Hormone release, transport, biological activity and breakdown can vary considerably and should be taken into account when evaluating an impact upon target tissues.
Endocrine activity during exercise	During exercise many hormones are released and a variety of cells are targeted for a range of different responses. Important responses under hormonal control relate to cardiovascular regulation and metabolic activity. Exercise intensity and duration can exert specific but different effects in a range of hormones.
Endocrine adaptations to training	There are less research data available that document the response of hormones to training. The evidence available is primarily related to endurance training and this seemingly blunts the hormonal response to acute exercise.

Related topics

Energy sources and exercise (B1)
Control of energy sources (B4)
Energy for various exercise
  intensities (B5)
Responses to training (B6)
Muscle structure (C1)
Motor-neural control of contraction
  and relaxation (C2)

Pulmonary adaptations to exercise
  (D)
Cardiovascular adaptations to
  exercise (E)
Exercise and environmental stress
  (I)

Role of the endocrine system

The endocrine system supports and interacts with the neural system to promote control and homeostasis with the human body. By the nature of its organization and activation the endocrine system is much slower than the neural system. As such it is less important for the initial activation and functional response to exercise, such as heart rate acceleration that strives to maintain homeostasis. The endocrine system has important roles to play in managing and controlling metabolic activity and cardio-respiratory function over prolonged periods of exercise and into the recovery period.

Structure of the endocrine system

The definition of a hormone is a chemical substance released from a cell or gland that is transported by fluids within the body to a target cell where a physiological effect will occur. Thus the endocrine system is composed of the tissues that produce and release the hormones, the hormones themselves, and the transport medium and target cells for the hormones. There are many different hormones, not all of which can be covered within the scope of this text, so readers are directed to other endocrine-based texts for further information. Here we will specifically concentrate on those hormones with a clear and defined role in maintaining homeostasis during exercise.

A number of tissues produce hormones but most are considered endocrine glands (see *Table 1*) and include the pancreas, the hypothalamus, the pituitary gland and the adrenal glands amongst others. These glands secrete hormones into the bloodstream through which they are transported to their target cells. Other tissues such as the heart, the kidney, the liver, endothelial cells and adipose tissue secrete hormones into extracellular fluid and then move to nearby cells. There are many different endocrine glands and tissues producing

Table 1.   *Major hormones involved in regulating the response to exercise*

Hormone	Site release	Relevance to exercise, training and health
Adrenaline (and Noradrenaline)	Adrenal Medulla	Regulation of metabolism and cardiovascular control
Adrenocorticotrophic Hormone (ACTH)	Pituitary (Anterior)	Regulation of hormone release from adrenal cortex
Aldosterone	Adrenal Cortex	Regulation of electrolyte and fluid balance in kidney
Anti-diuretic Hormone (ADH)	Pituitary (Posterior)	Regulation of fluid balance in kidney and cardiovascular control
Calcitonin	Thyroid	Regulation of circulating calcium levels
Cortisol	Adrenal Cortex	Regulation of metabolism, muscle catabolism, immune function role
Erythropoietin	Kidney	Regulation of production of red blood cells
Estrogen	Ovaries	Regulation of fat metabolism and bone turnover
Follicle Stimulating Hormone (FSH)	Pituitary (Posterior)	Regulation of sex hormone secretion
Glucagon	Pancreas	Regulation of metabolism via blood glucose control
Growth Hormone	Pituitary (Anterior)	Regulation of metabolism, bone turnover, muscle anabolism; controls release of IGF
Hypothalamic Release Hormone	Hypothalamus	Stimulation and inhibition of release of pituitary hormones
Insulin	Pancreas	Regulation of metabolism via blood glucose control
Insulin-like Growth Factor (IGF)	Liver	Regulation of anabolic activity in muscle
Luteinizing Hormone (LH)	Pituitary (Posterior)	Regulation of sex hormone secretion
Parathyroid Hormone	Parathyroid	Regulation of calcium and phosphate levels
Progesterone	Ovaries	Regulation of catabolic activity in muscle tissue and bone turnover
Testosterone	Testes	Regulation of anabolic activity in muscle tissue and bone turnover
Thyroid Stimulating Hormone (TSH)	Pituitary (Anterior)	Regulation of release of hormones from the thyroid gland
Thyroxine	Thyroid	Regulation of metabolic rate

hormones and indeed a vast array of hormones themselves controlling many aspects of the human response to exercise.

As noted in Section B4 hormones are categorized as either amine, peptide or steroid hormones. Amine hormones are small molecules formed from the amino acid tyrosine and include the catecholamines adrenaline (epinephrine) and noradrenaline (norepinephrine), both of which fulfill roles as hormones and neuro-transmitters. Peptide hormones are peptides consisting of between three and over 200 amino acids and include all the hormones released from the hypothalamus, pituitary, and pancreas (insulin and glucagon). Steroid hormones are lipid-based (cholesterol) and include all the hormones released from reproductive organs (testosterone and estrogen) and the adrenal cortex (aldosterone and cortisol).

Hormones are complex in that some have very specific target cells and very specific outcomes. Others, such as cortisol, may target a range of different cells (skeletal muscle and adipose tissue) and have very different effects at each site. In many instances hormones are seen to work together (both adrenaline (epinephrine) and cortisol promote free fatty acid release from adipose tissue) or they can work against each other (glucagon and insulin have opposing effects upon glucose uptake by target cells).

Activation and action of the endocrine system

The release of hormones can be promoted by neural, hormonal or other blood-borne stimuli. With neural stimulation a neurotransmitter is released that will activate the hormone-producing tissue. An obvious example is when sympathetic neural stimulation results in the release of adrenaline (epinephrine) and noradrenaline (norepinephrine) from the adrenal glands. An example of where a hormone, sometimes referred to as a hormone-releasing factor or trophic hormone, stimulates the secretion of a second hormone is provided by the control of cortisol release. Cortisol is released from the adrenal cortex under the action of adreno-corticotrophic hormone that is released from the anterior pituitary gland, which is in turn released in the presence of corticotrophin-releasing hormone produced by the hypothalamus. Finally hormones can also be released in response to some bloodborne stimuli and a good example is when the pancreas releases insulin upon the detection of high levels of blood glucose.

The nature of hormone release varies considerably. Some hormones, like adrenaline (epinephrine) can be released quickly when demanded in response to specific neural stimuli. Others are released in some repetitive pattern over a day or even over a month, as the female sex hormones demonstrate, while others are released in a random fashion. Associated with this fact is that the rate of release also varies as will its removal. Many hormones also travel in the circulatory system bound to proteins (e.g. testosterone) as well as in an unbound (free) state and this will alter the biological effect of the hormone, as bound hormones cannot interact with receptor sites. The combination of these factors will result in different kinetics of action upon the target cell. Some hormones act very quickly (especially those with high unbound concentrations such as the catecholamines) and some take prolonged periods of time to have an effect.

Once hormones are released they will effect their action after binding to receptor sites on the target cell. The number of receptors, the delivery rate of unbound hormone and the affinity of the bond between hormone and receptor limit receptor activation. Receptors themselves are not static entities and in some cases receptors can increase or decrease in number over short periods of time. The increase or decrease in receptor numbers often occurs in response to changing hormonal availability.

The process from hormone-receptor binding to the activation of the physiological response in the cell is also quite complex. One process, called second messenger activation, has been described previously in Section B4. In these examples amine and peptide hormones, which cannot diffuse through lipid-based cell membranes, exert their effect via activating another 'second' messenger within the cell. For example a hormone, via receptor binding, activates the enzyme adenyl cyclase, which transforms ATP to cyclic AMP (cAMP). cAMP, a very common second messenger, then sets off a catalog of metabolic reactions that eventually result in the desired cellular response. The activation of glycogenolysis in skeletal muscle cells under the action of adrenaline (epinephrine) is an example of second messenger activation (see Section B4).

Steroid hormones use a different process of activation. Steroid hormones are lipid-based and these can diffuse through the cell membrane. Once through the cell membrane and within the cell the steroid hormone will bind to specific receptors that then enter the nucleus. The hormone–receptor complex activates the transcription of specific genes. The resulting mRNA is translated to synthesize the proteins that bring about the hormonal response.

Endocrine activity during exercise

During exercise many hormones are released and a variety of cells are targeted for a range of different responses. Important responses under hormonal control relate to cardiovascular regulation and metabolic control.

The primary hormones that affect cardiovascular function are adrenaline (epinephrine) and noradrenaline (norepinephrine) although hormones such as antidiuretic hormone and aldosterone are secreted to promote fluid retention and thus maintenance of blood volume and blood pressure in a situation where fluid loss due to sweating is increased. Concentrations of adrenaline (epinephrine) and noradrenaline (norepinephrine) are both increased with exercise to maintain cardiac output, blood flow to the active muscles and blood pressure. We have previously stated that exercise intensity influences adrenaline (epinephrine) and noradrenaline (norepinephrine) release and it must be added that exercise duration (see *Fig. 1*) also alters their release.

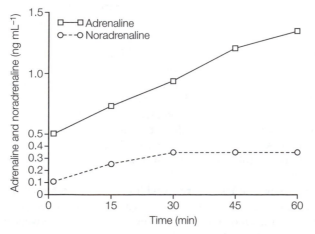

Fig. 1.   Adrenaline (epinephrine) and noradrenaline (norepinephrine) accumulation in the blood during prolonged exercise.

A range of hormones accomplishes regulation of metabolism, in a variety of target cells. The overall role of hormonal control is to make sure enough ATP is being produced to sustain muscle contraction and that blood glucose levels are maintained to supply the CNS with fuel. Important hormones, and their response to exercise, include an increase in glucagon (and a decrease in insulin which acts antagonistically to glucagon), adrenaline (epinephrine), noradrenaline (norepinephrine), growth hormone and cortisol (see *Table 1*). These hormones target adipose cells where fat mobilization is promoted (fat storage is inhibited), the liver where glycogen is broken down to glucose as well as skeletal muscle where breakdown of glycogen is promoted as is the uptake of free fatty acids (both to fuel metabolism of the contracting muscles and the recovery process).

As with neural activity the response of a variety of hormones implicated in metabolic and cardiovascular control differs dependent upon the selection of exercise mode, duration and intensity. These differences have received some research attention although we do not yet know all that is to be known. We have already noted that maximal exercise elicits a large increase in both adrenaline (epinephrine) and noradrenaline (norepinephrine). Data also suggest static and dynamic resistance exercises also result in elevated levels of these hormones. This is not too surprising given that all types of exercise will require an elevation in cardiovascular work. The response of hormones involved in metabolic regulation generally shows an increase with exercise intensity although the rise in insulin towards maximal exercise is somewhat surprising given its antagonistic role to fuel mobilization. Whilst both plasma growth hormone and cortisol concentrations increase with exercise intensity (see *Fig. 2*) it is interesting to note that at moderate exercise intensities a decline may occur with prolonged exercise duration. Both cortisol and growth hormone increase significantly with resistance exercise likely due to high exercise intensities employed. Despite such knowledge research work in this area is still far from complete.

Endocrine adaptations to training

There are less research data available that clearly document the response of a variety of hormones to different types of training regime. In the absence of such information a full mechanistic understanding of homeostatic control after training still eludes us.

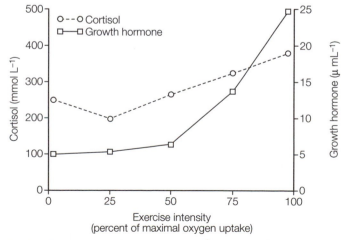

Fig. 2. Cortisol and growth hormone responses to graded exercise.

What evidence is available stems primarily from aerobic training programs and the response to prolonged exercise pre- and post-training.

Hormonal concentrations at rest are altered very little between pre- and post-training assessment. Exercise response data for hormones involved in cardiovascular and metabolic regulation suggest a widespread attenuation of hormone responses to exercise. For example noradrenaline (norepinephrine), adrenaline (epinephrine) and glucagon are all lower at the same absolute exercise intensity post-training (see *Fig. 3*). Linked to these changes is a reduction, post-training, in the drop in insulin concentration with exercise. The reduction in noradrenaline (norepinephrine) and adrenaline (epinephrine) are the consequence of reduced sympathetic drive post-training. Conversely, peak-exercise values for noradrenaline (norepinephrine) and adrenaline (epinephrine) seem to be augmented post-training but probably reflect the greater exercise capacity attained.

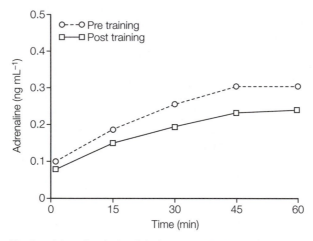

Fig. 3.   *Adrenaline (epinephrine) response to an acute bout of endurance running pre- and post-training.*

G1 FATIGUE AND ERGOGENIC AIDS

Key Notes

Ergogenic aids	An ergogenic aid is a food or food component which purports to improve the capability of an athlete to improve their performance. Foods include the macronutrients and fluids, whereas food components include naturally occurring substances found in the foods.
Fatigue and energy for high-intensity exercise	Fatigue, the inability to sustain the necessary power output for high-intensity exercise is dependent on muscle concentrations of PCr, glycogen, and build up of lactic acid.
Fatigue and prolonged exercise	The likely causes of fatigue for prolonged exercise are complex because many factors including physiological aspects as well as psychology are possible contributory factors. However, from a physiological perspective, depletion of muscle glycogen, hypoglycemia, and dehydration are three major considerations.
Related topics	Energy sources and exercise (B1) The endocrine system (F2)
	Changing body mass and body fat (J3)

Ergogenic aids

In their desire to succeed, athletes have resorted to ways of boosting performance by using physiological, nutritional and pharmacological agents. Such agents, which purport to improve an athlete's performance, are known collectively as **ergogenic aids**. The spectrum of ergogenic aids includes normal foods consumed at one end with drugs at the other end. In between these extremes, there are a range of nutritional products which normally occur in everyday foods at low levels, but can be taken in higher doses as supplements because of purification or synthesis processes.

In order to be recognized as an ergogenic aid, the nutritional product should have a recognized theoretical effect. Consequently, the likely benefits of such an aid require examination of the likely causes of **fatigue** during exercise and whether these are positively influenced by the product ingested.

Fatigue and high-intensity exercise

Section B briefly described factors likely to result in fatigue during exercise of varying intensities. Factors likely to influence fatigue during high-intensity exercise are those in which depletion of an energy source may present a problem or those in which accumulation of an energy product may cause the problem. The sources producing energy for high-intensity exercise are those involving the **PCr** system and **glycolysis** in the main. Potentially a decrease in PCr or muscle glycogen content could prove problematical, although equally an elevation of **muscle lactic acid** could prove disadvantageous.

Is there any evidence that PCr is depleted during intense bouts of exercise? In either single bouts of intense exercise lasting for 30 to 45 seconds or in multiple bouts of short duration high-intensity exercise or electrical stimulation of muscle, a total depletion of PCr may be observed. Furthermore, the reduction in PCr content of muscle correlates with a diminished power output. In the last 10 years, numerous studies have explored the benefits of supplementing the diet with **creatine** in an attempt to boost muscle PCr content and thus improve performance. The use of creatine supplementation is the focus of Section G4.

Another energy source whose depletion may cause fatigue during high-intensity exercise is that of muscle glycogen. **Muscle glycogen** is the carbohydrate source for anaerobic glycolysis, a key energy-producing process during intense exercise. The glycogen content of **fast glycolytic (FG) fibers** is of particular importance during such exercise. Although there is no evidence that elevated levels of muscle glycogen promote intense activity, there is evidence that low levels of muscle glycogen prior to such exercise may prove disadvantageous. In a series of experiments performed in the late 1980s, Greenhaff and colleagues showed that whereas a high carbohydrate diet failed to enhance cycling at 120% $\dot{V}O_{2max}$, a low carbohydrate diet resulted in an attenuated performance (see *Fig. 1*).

In spite of evidence that depletion of stores of muscle PCr and glycogen can impair performance at high-intensity exercise, there is a considerable amount of data on the fact that sprint activities lead to a significant increase in muscle lactic acid concentrations. The pH of muscle cells may change from a value of around 7.0 to that of 6.4 following intense bouts of exercise. Furthermore, it has been shown that if **muscle pH** decreases to a value of 6.4 there is impairment of key regulatory enzyme activity. The enzymes **glycogen phosphorylase** (which produces glucose from glycogen), **PFK** (the key regulatory enzyme for glycolysis), and even the **ATPases** may be inhibited. Such a situation would invariably lead to slower energy production and an attenuated power output resulting in fatigue. It should be remembered that when lactic acid is produced, it dissociates into lactate ions and hydrogen ions:

$$HLa \leftrightarrow La^- + H^+$$

It is the **hydrogen ions (H$^+$)** that bring about the decrease in pH as well as competing with **calcium ions (Ca^{2+})** for binding sites on the **crossbridges**. Hence the possibility that accumulation of lactic acid in muscle may contribute to fatigue. Section G5 explores the potential for alkalinizers, such as sodium

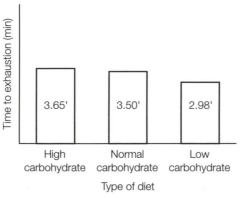

Fig. 1. Effects of a high carbohydrate, normal and low carbohydrate diet on time to exhaustion at 100% $\dot{V}O_{2max}$.

bicarbonate and sodium citrate, as ergogenic aids which may affect the influence of lactic acid on muscle.

Fatigue and prolonged exercise

As with factors likely to cause fatigue during high-intensity exercise, those concerned with prolonged exercise are complex and interrelated. The major physiological factors are those of depletion of muscle and liver glycogen stores, with the latter resulting in **hypoglycemia**, as well as overheating due to dehydration. Of course there are psychological factors which are important, but that debate is beyond the scope of this book.

Lipids are a major and significant source of energy for prolonged activities, although they can only contribute to approximately 100% of the supply if exercise intensity is around 50% $\dot{V}O_{2max}$ or lower. If exercise is to be conducted at intensities greater than 50% $\dot{V}O_{2max}$, then the importance of carbohydrates is realized. Unfortunately the total carbohydrate stores in the body are limited and may last for 70 to 90 minutes at 65–70% $\dot{V}O_{2max}$. The evidence for this is provided in the next section (i.e. G2).

During prolonged exercise the muscles produce heat, which has to be dissipated. This is achieved by promoting **sweating** and **vasodilation** of blood vessels at the periphery. Evaporative sweat loss enables, via conduction, the blood to be cooled. The net result is that blood returning to the body core is cooled. However the loss of sweat diminishes the fluid reserves of the body, and if left unchecked, results in **dehydration**. If body fluid loss is relatively severe, the heat loss mechanism shuts down in an attempt to conserve body water. The result is that the body temperature becomes elevated above safe levels and **heat stroke** may be the consequence. Section G3 briefly addresses the findings in relation to the relevance of fluid balance in dehydration and **rehydration**.

G2 MACRONUTRIENTS

Key Notes

Carbohydrates	Carbohydrates are an essential macronutrient for muscles to function effectively at exercise intensities above 50% $\dot{V}O_{2max}$. They are stored in muscle and liver as glycogen, although blood glucose is also an available carbohydrate energy source.
Lipids	Lipids are a major source of energy for prolonged exercise but can only be used at relatively low levels of exercise intensity. Stimulating fatty acid release from lipid stores could spare the limited muscle glycogen stores and so be of ergogenic benefit to athletes.
Protein and amino acids	Muscles are made up of proteins and amino acids. During exercise there is a breakdown of muscle and liver protein to produce amino acids, which may then be used either directly or indirectly as an energy source. Breakdown of protein can be limited by ingesting carbohydrates before or during exercise.
Related topics	Bioenergetics for movement (B) Energy balance, body composition and health (J)

Carbohydrates

The body stores carbohydrate as glycogen in muscle and liver essentially. The total amount may be approximately 300–400 grams, although this is affected by diet and exercise. The importance of carbohydrates for sport performance was recognized at least as long ago as the early 1920s when a group of scientists examined factors contributing to fatigue in the Boston marathon. They discovered that some subjects finished in a very poor physical and mental state, and that this was because they were **hypoglycemic**, i.e. had a blood glucose concentration lower than 4 mM. In the following year when these runners consumed carbohydrate before and during the marathon they finished in a better state and were not hypoglycemic.

In the 1930s a series of studies in Scandinavia once again highlighted the role of carbohydrates as being crucial for prolonged exercise, and although hypoglycemia was again implicated in some instances (see *Fig. 5* in Section B5), others subjects became fatigued in spite of possessing normal or even slightly elevated blood glucose concentrations. The conclusions were that a high carbohydrate diet in the days before exercise or carbohydrate ingestion during prolonged exercise enhanced performance.

It was not until the late 1960s, after the advent of the **needle biopsy** technique, that the importance of muscle glycogen was realized. The classical work of the Scandinavian scientists clearly demonstrated that there was a relationship between carbohydrate intake in the days before exercise led to an increase in **muscle glycogen** content, which in turn promoted time to exhaustion on a cycle ergometer (see *Fig. 4* in Section B5).

The muscle biopsy technique became a powerful tool in the armory of sport physiologists throughout the next three decades. The relationship between exercise intensity and muscle glycogen use (*Fig. 1*) again highlighted the relevance of muscle carbohydrate stores for exercise of reasonable intensities.

Since **muscle fibers** exist as **FG**, **FOG**, and **SO** types, the question arises as to whether muscle glycogen content is evenly depleted during exercise of varying intensities. A study performed in Scandinavia demonstrated that when exercise was approximately 45% $\dot{V}O_{2max}$ the glycogen stores in all fibers were very slowly utilized but not depleted even after 4 hours, at 65% $\dot{V}O_{2max}$ the SO, FOG and FG fibers were all depleted at the point of fatigue (i.e. 120 minutes), whilst at 90% $\dot{V}O_{2max}$ only the FG fibers were totally depleted (*Fig. 2*). This study demonstrated the relationship between muscle glycogen depletion, recruitment of muscle fibers and exercise intensity.

A 'field-based' study, which also highlighted the value of muscle glycogen for sports performance, was undertaken on Swedish soccer players (see Section B5 and *Table 1* in Section B5).

The 1970s and 1980s saw a plethora of studies that highlighted the importance of carbohydrate feedings to aid prolonged exercise. These studies employed a range of carbohydrate sources including **glucose**, **fructose**, **maltose**,

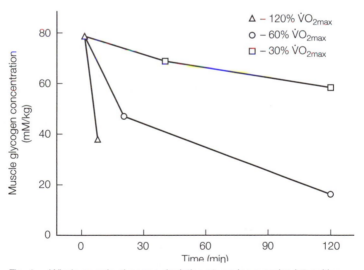

Fig. 1. *Whole muscle glycogen depletion at varying exercise intensities.*

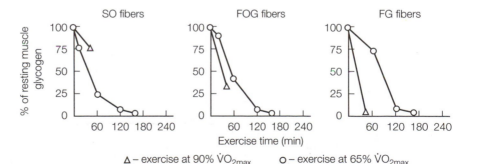

Fig. 2. *Muscle glycogen depletion in varying muscle fiber types at varying exercise intensities.*

sucrose, **glucose polymers** (**maltodextrins**), and even starch. With the exception of fructose, all carbohydrates ingested during exercise generally enhanced performance. Furthermore, the findings from studies employing **stable isotopes** or **radio-isotopes** of the carbohydrate source found oxidation rates of the ingested carbohydrate source did not exceed 60 g h^{-1}. The proposed advantage of consuming carbohydrates immediately before or during exercise is that the ingested (and oxidized) carbohydrate spares the limited muscle glycogen stores and so offsets fatigue. *Fig. 3* illustrates the effect of cyclists ingesting a 6% carbohydrate drink during exercise to fatigue at 70% $\dot{V}O_{2max}$. These elite cyclists were able to exercise for a further 60 minutes when carbohydrate was ingested compared to a placebo drink.

Clearly carbohydrates are an important fuel for prolonged exercise and sports performance. The questions remain as to what duration of exercise and at what intensities of exercise are carbohydrates unlikely to prove ergogenic. Research has demonstrated that relatively intense bouts of activity less than 60 minutes in duration are unlikely to benefit from carbohydrate ingestion immediately before or during that exercise. This is true as long as there is a reasonable store of muscle glycogen prior to that exercise, since if the levels are low then carbohydrate ingestion will be useful. Therefore carbohydrate ingestion during exercise is of importance if the exercise is greater than 60 minutes in duration. Furthermore, if the exercise intensity is around 50% $\dot{V}O_{2max}$ or less then carbohydrate feeding is unlikely to prove useful since fats can provide a significant proportion of the energy.

Very intense bouts of exercise can only be sustained for short periods of time. Can carbohydrates be useful as an energy source for these types of activities? Results from a study reported in the late 1980s in which subjects ate a high, low or normal carbohydrate diet for 3 days and then undertook cycling to fatigue at an exercise intensity corresponding to 100% $\dot{V}O_{2max}$ showed that whereas the high carbohydrate diet did not significantly improve performance compared

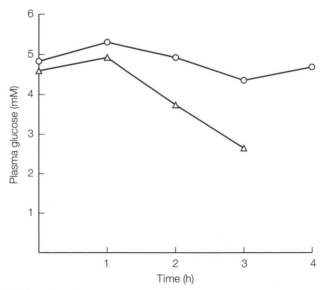

Fig. 3.   Blood glucose concentration and time to fatigue when exercising with ingestion of a carbohydrate drink (○) or placebo (△).

with the normal carbohydrate diet, the low carbohydrate diet resulted in an attenuated performance. More recently, it has been shown that taking carbohydrate drinks in between bouts of high-intensity exercise can sustain exercise at higher intensities than plain water.

Carbohydrates are clearly a major source of energy for intense as well as for moderate intensities of exercise. It would seem pertinent that ensuring muscle carbohydrates are rapidly restored after exercise is important. A series of studies in the late 1980s demonstrated that carbohydrates should be ingested within an hour of completing the exercise, that the amount of carbohydrate should be approximately 2 g kg^{-1} body mass in the first 2 hours, and that the form of carbohydrate is not wholly relevant (although simple sugars may be preferable in the first instance).

Lipids

Triglyceride stores in **adipose tissue** and muscle are important sources of energy for prolonged exercise, although lipids cannot be used during intense bouts. The triglyceride stores have sufficient energy stored for hours and even days of activity, and are unlikely to be depleted. Therefore it appears that consideration for lipids and fats is not necessary for performance. However, it needs to be re-stated that lipid sources take some time before they are mobilized and used, and that in the intervening period carbohydrates will be used. Furthermore, lipids are unable to be used at relatively high exercise intensities although endurance training enhances this capability. So if fats can be mobilized earlier in exercise or their use promoted, there will be inevitable sparing of the limited muscle glycogen stores. Section G4 explores the potential ergogenic aids which may stimulate fatty acid release (e.g. **caffeine**), provide a possible alternative fast-uptake lipid source (e.g. **MCT**), or hasten fatty acid uptake into the **mitochondria** (e.g. **carnitine**). Under these circumstances lipids may prove a useful energy source.

Conversely, body fat percentages of athletes should be maintained at relatively low levels since the extra weight may be inefficient in energetic terms. Male and female athletes should possess % body fat scores of no more than 15% and 25% respectively. The average population range for males and females not considered **obese** are 10–20% and 20–30% respectively. Athletes may therefore engage in 'fat-burning' strategies at some times in order to reduce **body fat** to acceptable limits. This can be achieved by a regime of training (see Section H) and diet, including the use of some dietary agents which purport to stimulate **fat oxidation**.

Protein and amino acids

Muscles, **enzymes**, cell membranes, **myoglobin, hemoglobin, albumin**, and some key **hormones** are all made up of proteins, and hence the need to ensure adequate protein intake for health. But is there an extra requirement for athletes? The World Health Organization has suggested a daily recommended intake for protein of 0.8 g kg^{-1} body mass for adults. The requirement for athletes is approximately 1.4–1.8 g kg^{-1} body mass. This is because **protein degradation** is an inevitable consequence of exercise, although the amount of degradation can be attenuated by ensuring adequate carbohydrate intake during exercise, i.e. carbohydrate spares protein breakdown.

The requisite amount of protein needed can be estimated by what is known as **nitrogen balance**. In essence this technique determines the amount of nitrogen (protein) in the diet and then assesses that against the nitrogen loss from feces, urea, and sweat urea. If the diet contains overall greater amounts of nitrogen than that lost, the body is in **positive nitrogen balance**, and if there is greater loss than intake the body is in **negative nitrogen balance**. Balance occurs when the net loss and intake are equal. *Fig. 4* shows the relationship between nitrogen balance and protein intake.

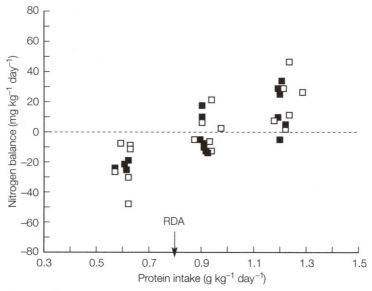

Fig. 4.   Nitrogen balance and daily protein intake.

Negative nitrogen balance is not desirable for athletes nor for sedentary individuals. Athletes should therefore ensure that their protein intake is around 1.4–1.8 g kg^{-1} body mass per day, and even consider increasing this to 2.0 g kg^{-1} body mass per day for a period of 7 days if training load is increased. Having a very high protein intake for prolonged periods of time is unnecessary and may also be harmful. The end product of protein metabolism is **urea**, which is removed via the kidney. It is possible to place the kidney under stress if the diet is too high in protein, especially if fluid intake is compromised.

Recommendations to athletes undergoing rigorous training as well as those wishing to increase muscle mass would be to ensure the guidelines above are followed in relation to total protein intake but also to consider ingesting around 10–15 g of easily digestible protein (i.e. whey protein, milk or milk shake) an hour before exercise and then a further 10–15 g within an hour after exercise. Such a consideration should minimize protein degradation and promote **protein synthesis** in recovery.

Amino acids are mobilized from protein during exercise, particularly in a carbohydrate-depleted condition. This is because the **amino acids** are converted to **alanine** in muscle, which is then transported to the liver where it is converted to glucose. In other words the amino acid, alanine, can be used to make glucose (*Fig. 5*). This process, the **glucose–alanine cycle**, is diminished if plenty of carbohydrates are available during exercise. Hence the value of carbohydrate feeding during exercise.

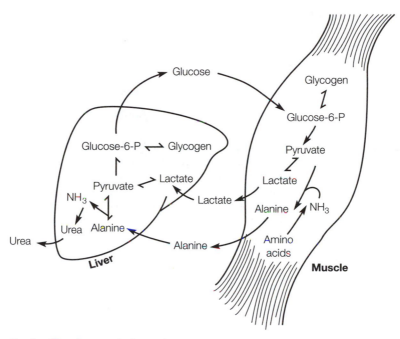

Fig. 5.   The glucose–alanine cycle.

G3 FLUIDS

Key Notes

Dehydration	When water is lost from the body either via sweating but also through urine excretion, and is not replenished by drinking, the total body fluid loss leads to dehydration. Essentially this is a loss of total body water. Dehydration invariably leads to diminished stamina, power, speed, and strength. The greater the level of dehydration, the greater the loss of muscle function. Significant losses of body fluids may result in heat stress, heat stroke, and even death.
Rehydration before and during exercise	The process of rehydration involves replacing the body water loss by ingesting water or other fluids. During prolonged exercise (especially in the heat) there is a need to imbibe around 200 ml every 15–20 minutes in an attempt to rehydrate. In many cases even this amount is not enough to prevent dehydration.
Rehydration after exercise	Total rehydration is possible after exercise. The volume of fluid should be 1.5× greater than the total fluid loss, and the fluid should contain sodium.
Related topics	Cardiovascular responses to exercise (E3) Thermoregulation (I1) Exercise in hot and humid environments (I2)

Dehydration

Heat generated by the muscles during exercise must be dissipated in an attempt to maintain body temperature within a narrow physiological band. **Evaporative heat loss** via **sweating** is a major means employed by the body to cool down. Sweating does not only occur at high ambient temperatures but also happens when exercising in a cool environment at a high exercise intensity. Under such circumstances, sweat rates between 1 and 3 L h^{-1} are not uncommon. **Body mass losses** in marathon runners range from 1–4 kg (i.e. 1–6% loss for a 70 kg person) at 10°C to more than 5.5 kg (8% loss) in warmer environments.

A body water loss of as little as 1% of body mass impairs performance whilst a 4% loss impairs exercise capacity by as much as 30%. **Heat exhaustion** can occur at around 5% body mass loss and circulatory collapse and heat stroke at around 10% body mass loss due to water loss (see *Fig. 1*).

Rehydration before and during exercise

If sweat losses of at least 1–2 L h^{-1} are experienced while performing low-intensity exercise, is it possible to maintain **euhydration** during exercise by drinking? Thirst is not a good indicator of body water requirements since it is not perceived until an individual has incurred a water deficit of around 2% body weight loss. Therefore studies which rely on *ad libitum* intake of water, result in incomplete replacement of body water losses and invariably lead to some level of dehydration. Forcing subjects to drink can prevent dehydration as long as the sweat loss is not too great, i.e. approximately 1 L h^{-1}. It is difficult and

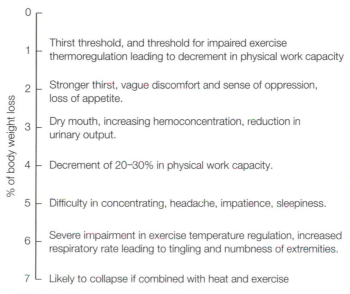

Fig. 1.   Effects of fluid loss on body functions.

uncomfortable to drink more than 1 L h^{-1}, and this is normally achieved by regular ingestion over that time i.e. 150–200 ml every 10–15 min.

It is desirable for those who are engaged in exercise to become hydrated before starting the bout. This can be achieved by drinking fluids with the meal before the exercise, and also ingesting 300–500 ml of fluid immediately before the exercise.

What kind of fluid should be given? It is clear from numerous studies that a **carbohydrate-electrolyte** solution helps to maintain blood volume, assists thermoregulation, reduces the risk of heat injury, provides exogenous energy, and thus enhances performance during prolonged exercise. However, in order to prove beneficial, the drink should be capable of being emptied from the stomach (**gastric emptying**) rapidly and then **absorbed** across the **intestine** rapidly. The rate of gastric emptying is affected by numerous factors such as exercise intensity, as well as concentration, **osmolality**, and temperature of the drink. Exercise intensities greater than 65% $\dot{V}O_{2max}$ impair gastric emptying as does an increase in concentration and increase in temperature of the ingested solution. Pure, cool water would thereby empty at a greater rate than a concentrated, warm glucose drink.

If the major factor to offset fatigue is dehydration, then water would be the optimal drink. If however the major concern is maintenance of carbohydrate levels, then a carbohydrate-electrolyte drink would be preferable. The various **isotonic** sports drinks have been produced in an attempt to address the issue concerning rehydration and energy provision. *Fig. 2* illustrates the fact that as there is an increase in the glucose content of a drink there is a greater provision of carbohydrates but an attenuation of water availability. The compromise appears to occur at a glucose concentration of around 6%, since at concentrations above this value there is a significant negative impact on water delivery.

A 6% glucose solution is likely to have an **osmolarity** of around 280 mOsm L^{-1} which is isotonic with body fluids (which have an osmolarity of around

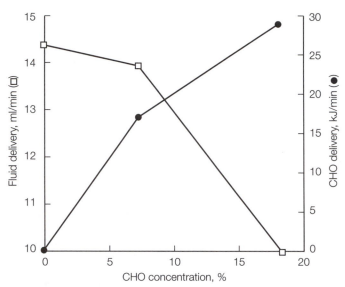

Fig. 2. Relationship between increasing glucose concentration of a drink and the availability of water and carbohydrate.

290–300 mOsm L^{-1}). As the concentration of dissolved material in the drink increases, so does the osmolarity. Osmolarity is a measure of the osmotic pressure exerted by dissolved particles in a fluid. The greater number of dissolved particles the greater the osmolarity. So a 6% glucose solution has a lower osmolarity than a 10% glucose solution, and a 6% glucose solution has a lower osmolarity than a 6% glucose-electrolyte solution (because the electrolytes add to the osmolarity). Therefore drinks can be classified as **hypotonic** (<270 mOsm L^{-1}), isotonic (270–300 mOsm L^{-1}), or **hypertonic** (>330 mOsm L^{-1}). *Table 1* highlights osmolarities and concentrations of carbohydrates in selected drinks.

The addition of the electrolyte, sodium, aids gastric emptying and **intestinal absorption** of both water and glucose. It must be remembered that glucose is co-transported with sodium across the intestinal lumen wall, and that water is thereby 'dragged' across the membrane passively. The faster the rate of movement of glucose across the membrane, the faster the rate of water movement;

Table 1. Carbohydrate concentration and osmolarity of selected drinks.

Drink	Carbohydrate concentration (%)	Osmolarity (mOsm/L)
Coca Cola	11.0	650
Fanta	10.8	478
Sprite	11.0	590
Red Bull	10.7	686
Fresh orange juice	10.0	670
Lucozade	19.2	900
Lucozade Sport	6.0	290
Gatorade	6.0	350
Powerade	8.0	300
Isostar	7.6	305
Exceed	7.0	270

the more **sodium** in the solution, the faster the movement of sodium and glucose. Hence sports drinks are often known as carbohydrate-electrolyte drinks because of the fact that sodium is normally present as the electrolyte. The amount of sodium is usually rather low in order to make the drink palatable, and although an ideal content of about 1 g L^{-1} would be desirable, the actual amounts in most sports drinks is rather less.

Rehydration after exercise

If sweat loss is significantly greater than 1 L h^{-1}, dehydration is inevitable since the volume likely to be comfortably ingested is around 1 L h^{-1}. The first opportunity to rehydrate is immediately after exercise. This is particularly important in tournaments when there is a relatively short recovery period between bouts. The rules of thumb for **rehydration** after exercise are to consume an amount of fluid 1.5 times greater than the fluid loss undergone (i.e. 1.5 L imbibed after a 1 L sweat loss), for the drink to contain sodium as an electrolyte (up to 1 g L^{-1}), and not to worry if there is a significant carbohydrate content in the drink. A player should be weighed before and after exercise, where the body weight loss in kg is equivalent to the water loss in liters.

The importance of sodium in the drink is to ensure that the fluid ingested is not excreted via urine. It must be remembered that the control of **water excretion** by the kidney is regulated by the hormone **ADH**, secreted from the **posterior pituitary** gland. Receptors in the posterior pituitary are able to detect changes in plasma volume and the sodium content of blood. If the plasma volume is increased, sodium content of the blood is diluted and so ADH is not released. The consequences are that the kidney fails to reabsorb some of the water passing through, and produces a copious, dilute urine. This happens when plain water is ingested. If the drink contains sodium however, then at least the sodium concentration of the blood is not compromised and some ADH may be secreted to conserve body water. *Fig. 3* shows how an increase in sodium content of a drink ingested after dehydrating exercise affects fluid balance after 6 hours recovery.

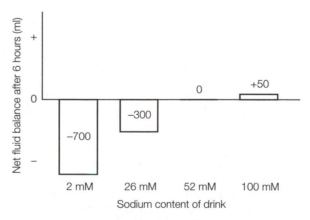

Fig. 3. Net fluid balance after 6 hours of recovery following ingestion of varying concentrations of sodium.

G4 NUTRITIONAL ERGOGENIC AIDS

Key Notes

Creatine	Creatine is an amino acid found in skeletal and cardiac muscle both in the free form (Cr) and in the phosphorylated form, creatine phosphate or phosphocreatine (PCr). PCr plays an integral role in rapid production of ATP from ADP during intense bouts of exercise.
Alkalinizers	Alkalinizers is the generic name given to compounds which have an ability to increase the alkalinity of blood and body fluids. Examples include sodium bicarbonate and sodium citrate. Since lactic acid production is an inevitable result of glycolytic activity during sprinting, alkalinizers may play a part in addressing the likely changes in muscle pH from lactic acid accumulation.
Caffeine	Caffeine is a naturally occurring compound found in many food and drink products. It is most widely used as a stimulant, and a major worldwide source is coffee. Caffeine may benefit athletes not only due to its stimulant properties but also because it stimulates release of fatty acids from adipose tissue.
Carnitine	Carnitine is a compound found in the membrane of mitochondria and is involved in the transport of long-chain fatty acids from the cytoplasm into the matrix of the mitochondria. Since the rate of entry of fatty acids into the mitochondria is a potential rate-limiting stage in the use of fatty acids as energy sources for prolonged exercise, there may be consideration to enhance mitochondrial carnitine levels and so promote fat oxidation.
MCTs	Medium-chain triglycerides (MCTs) are artificially produced compounds in which medium chain fatty acids are attached to a glycerol molecule. These compounds are more easily digested and rapidly absorbed across the intestine than long-chain fatty acids and could therefore be a potential useful lipid source of energy. In addition, medium-chain fatty acids can pass across the mitochondrial membrane without the need for carnitine transport. This more rapid uptake by the mitochondria may also be of benefit to athletes as an energy source during prolonged exercise.
Glutamine	Glutamine is the most abundant amino acid in the blood and also in the body's amino acid pool. It has been stated that appropriate levels of glutamine are required for optimal immune function.
Antioxidants	Intense or prolonged exercise demands a significant oxygen uptake by the body. Such an increase in oxygen consumption leads to an enhanced production of free radicals and concomitant related cell damage. The body uses antioxidants as a means of quenching free radicals and the damage they cause (notably to membranes). Antioxidants in foods include vitamins C and E as well as selenium. These compounds may protect muscle from damage during exercise.
Related topics	Rates of energy production (B2) Fiber types (C4) Energy stores (B3)

Creatine

Creatine is probably the most widely used of the nutritional ergogenic aids, and over the last 10 years also the most widely studied. **Creatine** has been reported to have been used by many successful athletes from a range of diverse sports. The use of creatine supplementation for sport originated in the early 1990s due to the fact that it may reduce muscle fatigue, because its phosphorylated form, **PCr**, can rapidly regenerate the ATP being used. If the muscle stores of Cr and PCr can be enhanced, then the possibility of being able to exercise more intensely before fatigue ensues is promoted. Furthermore recovery between successive bouts of intense exercise may be enhanced by the presence of greater concentrations of Cr and PCr.

The body synthesizes Cr from the amino acids **glycine**, **arginine**, and **methionine** at a rate of 1–2 g day^{-1}. This is undertaken by the liver from where it is transported to the muscle. An insulin-dependent transporter is present in muscle membrane to aid transport into the muscle cytoplasm. In addition to synthesis from amino acids, creatine can also be found in certain foods. These are notably fish and meat, putting vegetarians at a disadvantage in terms of creatine intake. The amount of creatine in fish and meat is relatively small and so the likely daily total intake and production of creatine is no more than 2–4 g day^{-1}. Since the daily creatine requirement is around 2 g day^{-1}, this can be met by synthesis and/or dietary intake.

Studies on resting muscle Cr and PCr content have shown values of about 50 mM kg^{-1} dry muscle and 75 mM kg^{-1} dry muscle respectively (i.e. 125 mM kg^{-1} dry muscle of total creatine), although immediately following repeated sprint bouts the Cr would be elevated and the PCr depleted. In order to possibly enhance intense exercise, would creatine supplementation increase Cr and PCr in muscle? If so, are there likely to be any positive, ergogenic effects?

Studies in the early 1990s demonstrated that feeding 20 g day^{-1} of creatine (taken as four 5 g portions through the day) for 5 days resulted in significant increases in total muscle creatine. Twenty-five percent of this total increase was due to an increase in PCr. However not all subjects responded to this form of supplementation, whereas all did so when supplementation was undertaken in conjunction with carbohydrate feeding. This pointed to the fact that insulin is involved in uptake of creatine by muscle. *Fig. 1* shows increases in total muscle creatine with supplementation, and highlights that there is an upper limit of muscle creatine which cannot be superseded. It appears that the maximal amount of creatine in muscle is unlikely to increase above a concentration of about 150 mM kg^{-1} dry muscle. This is an increase of around 30–40 mM kg^{-1} dry muscle from normal levels.

There are two phases to correct creatine supplementation. These are a loading phase of 5 days during which 20 g of creatine is ingested each day in four 5 g portions. This is followed by the maintenance phase during which 3–5 g of creatine are ingested each day to account for any likely loss. It must be stressed that because creatine is an amino acid, and that the end products of amino acid metabolism are excreted by the kidney, intake of water and other fluids is of paramount importance. Creatine supplementation and a low fluid intake or creatine supplementation in combination with diuretics is a potential major health hazard and must be avoided. Kidney damage could ensue if adequate drinking of fluids is not undertaken. Equally, taking 20 g of creatine a day for a long period also poses potential kidney problems, besides which such a strategy will not lead to any further increases in muscle creatine concentration.

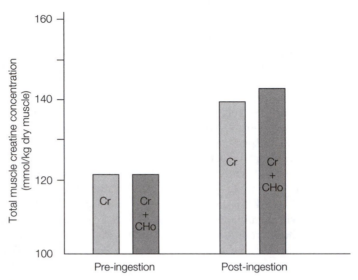

Fig. 1.   Changes in total muscle creatine (Cr + PCr) after five days of supplementation with creatine (Cr) or creatine with carbohydrate (Cr + CHo).

Can the increase in total muscle creatine improve performance, and if so what kind of activities are improved? Supplementation with creatine has been shown to improve repeated sprint or high-intensity activities, particularly if the bouts of activity are between 5 and 30 seconds. Longer bouts of repeated exercise or single bouts are generally not influenced. Furthermore, studies in which strength and power training are employed also generally result in gains in muscle mass, strength, and power. The effect of supplementation on performance in game simulations has not yet been elucidated, although there is likelihood that maintenance of sprint performance in intermittent bouts of activity such as in soccer is likely to happen.

Does creatine supplementation lead to an increase in lean body tissue (i.e. muscle mass rather than fat mass)? Any initial increase in body mass during the loading phase of creatine supplementation is likely to be due to an increase in body water. This is because creatine is an osmotically active substance which draws water into muscle cells when it is taken up. So the intracellular water content increases in muscle cells. If the supplementation is continued and strength and/or power training is undertaken, then muscle hypertrophy and increases in strength and power occur. Part of the reason for the increase in muscle mass is that muscle cells swell due to increased intracellular water, and the swelling stimulates protein synthesis. Therefore the increase in muscle mass following a period of supplementation and strength training is due to a combination of the training effect and enhanced protein synthesis from muscle cell swelling.

Alkalinizers

It is clear that a consequence of exercising at a high intensity is the production of lactic acid. The net effect of such a production is the accumulation of lactic acid in muscle and a decrease in pH from a normal resting value of 7.0 to values as low as 6.4. The resultant of this increase in acidity within the muscle is impaired enzyme activity and diminished energy (ATP) production. Alkalinizers, such as **sodium bicarbonate** and **sodium citrate**, increase the alkalinity of the blood (not the muscle), and so how can they affect muscle pH?

Lactic acid dissociates into lactate ions and hydrogen ions in solution. Once produced, lactate ions and hydrogen ions can diffuse out of the muscle into the blood along a pH gradient. Since the inside of a muscle has a pH of around 7.0 and **blood pH** is about 7.4, the gradient is sufficient to allow both ions to move outwards. If the blood pH is enhanced (i.e. made more alkalotic) by ingestion of alkalinizers, the potential to increase this pH gradient will result in a greater efflux of hydrogen and lactate ions from the muscle. Alkalinizers thereby influence the rate of efflux of lactate and hydrogen ions from the muscle rather than affect the pH of muscle *per se*. The consequence is that the inside of the muscle cell takes slightly longer to become acidotic and so offsets fatigue for a short period of time.

Sodium bicarbonate and sodium citrate have been shown to have positive effects on short-term, high-intensity exercise under the right circumstances. What are these circumstances? Firstly, sodium bicarbonate needs to be ingested at a dose of 0.3 g kg^{-1} body mass at least 2 hours before the exercise. Secondly, the intense bout of activity should be longer than 30 seconds but no longer than about 5 minutes since lactic acid accumulation is a likely causative factor for fatigue under these conditions. A potential problem with ingestion of sodium bicarbonate is that gastrointestinal problems may result. This can be alleviated if plenty of water is ingested when taking the dose and also if the dose is portioned rather than taken *in toto*. Sodium citrate has proved less problematic with regard to gastric problems, and so may be more useful.

Caffeine

Caffeine is a drug found to occur naturally in a wide variety of food components, and consequently has become socially accepted. Caffeine affects the central nervous system, muscle and adipose tissue, and hence may influence performance positively. There are three possible ways in which caffeine may have ergogenic properties. Firstly, caffeine may directly affect the central nervous system to improve perception of effort or affect **neural activation** of muscle. Secondly, caffeine may directly stimulate muscle **glycogenolysis** through activation of **glycogen phosphorylase** by release of calcium ions from the **sarcoplasmic reticulum**, and finally, caffeine may stimulate release of fatty acids from adipose tissue to be used as an energy source prior to and during exercise and so spare muscle glycogen in prolonged activities. The first two proposals relate to high-intensity exercise whereas the third relates to endurance performance.

For caffeine to have an ergogenic effect, the dose should be approximately 5–6 mg kg^{-1} body mass an hour before exercise. Any dose above 6 mg kg^{-1} body mass could result in some individuals exceeding the IOC regulations for caffeine in urine (although after January 2004 caffeine has been removed from the IOC list of banned substances). Furthermore, higher doses (i.e. above 6 mg kg^{-1} body mass) do not necessarily improve performance significantly above 5–6 mg kg^{-1} body mass. *Fig.* 2 illustrates the findings from a study in which subjects cycled to exhaustion following ingestion of caffeine at doses of 3, 6, and 9 mg kg^{-1} body mass compared with a placebo (0 mg caffeine). No significant advantage accrued above 6 mg kg^{-1} body mass.

Carnitine

Carnitine is marketed as a fat burner for athletes. This is because carnitine is required to transport long-chain fatty acids across the **inner mitochondrial membrane** into the matrix for β-**oxidation**, a proposed rate-limiting step in fat oxidation. Carnitine has also been shown to regulate the concentrations of acetyl-CoA and free CoA in the mitochondria by binding with the acetyl CoA and so increasing the availability of free CoA. An increase in CoA within the

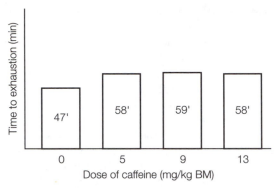

Fig. 2.   Effect of differing doses of caffeine on cycle time to exhaustion at 65% $\dot{V}O_{2max}$.

mitochondria promotes activity of the enzyme **pyruvate dehydrogenase** (PDH) and also aids the process of β-oxidation. The net effect is increased oxidation, an attenuated reliance on muscle glycogen, and a reduction in lactate production. *Fig. 3* is a schematic showing how carnitine fulfills these two roles.

Approximately 98% of the total stores of carnitine are found in skeletal and cardiac muscle. In healthy individuals the body is capable of synthesizing enough carnitine from the amino acids **methionine** and **lysine**. In addition, carnitine can be obtained via dietary sources such as meat, fish, and dairy products. Studies have shown that whereas oral supplementation with carnitine increases plasma concentrations, there is no significant uptake and incorpora-

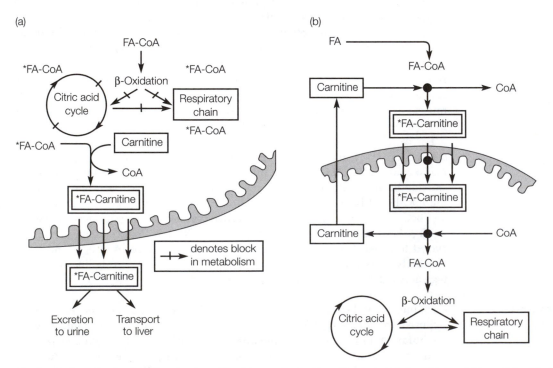

Fig. 3.   Schematic showing the role of carnitine (a) as a transporter of long-chain fatty acids, and (b) regulator of acetyl CoA and free CoA in the mitochondria.

tion into muscles. Does this imply that carnitine is unlikely to be of ergogenic benefit?

A recent review of 17 studies in which the potential ergogenic effects of carnitine were explored, showed that 13 of the studies found no effect on factors such as increases in $\dot{V}O_{2max}$, lactate concentration, or fat oxidation. Neither were there any significant improvements in performance measures. In these studies the carnitine was given at a dose of between 2 and 6 g day^{-1} for between 7 and 28 days. The four studies which found positive effects of carnitine all involved $\dot{V}O_{2max}$ and lactic acid as measures of ergogenic effect. Two studies gave the carnitine an hour before the test whilst the other two used periods of 14 and 28 days. It is difficult to rationalize why the studies whereby ingestion happened an hour before exercise should work. It appears that although there is a theoretical basis for an ergogenic benefit of carnitine supplementation, the overwhelming lack of support from studies reported contradicts the theory.

Medium chain triglycerides (MCTs)

MCTs are semisynthetic oil mixtures in which the long-chain fatty acids (LCT) attached to glycerol are replaced with medium-chain fatty acids, i.e. replace fatty acids which have more than 12 carbons with those which have between six to ten carbons. *Fig. 4* illustrates this point. As a result MCTs have characteristics different to LCTs. These include the fact that MCTs are liquid at room temperature, miscible with water, easily digestible, rapidly absorbed across the intestinal wall, transported to the liver by blood rather than lymphatic system, do not require carnitine for transport across the inner mitochondrial membrane and move in faster, and can be metabolized as rapidly as glucose in muscle. These characteristics have led researchers to explore the potential of MCTs as an exogenous energy source which would spare muscle glycogen and enhance endurance performance.

Over the last 20 years or so approximately ten well-controlled studies have been undertaken to examine the potential benefits of using MCTs either before or during endurance exercise. These studies have either used MCTs alone or in combination with carbohydrates, and have compared the effects on performance, fat oxidation, and muscle glycogen sparing. No studies reported an ergogenic benefit of MCT when taken singly, and only one study reported a positive effect of MCT when taken in combination with glucose (although a repeat study using the same protocol failed to show any positive effect). None of the studies found glycogen-sparing nor enhanced fat oxidation when MCTs were ingested. Furthermore, gastrointestinal problems were encountered when

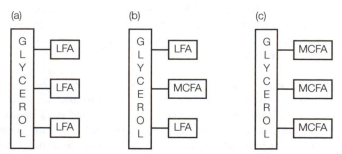

Fig. 4. Triglycerides with either all long-chain fatty acids (a), a mixture of long- and medium-chain fatty acids (b), or medium-chain fatty acids only (c).

the dose of MCT exceeded 30 g. It thus appears that MCTs are not a useful endogenous energy source during prolonged exercise.

Glutamine

Glutamine is the most abundant amino acid in plasma as well as skeletal muscle, and accounts for around 60% of the total intramuscular free amino acid pool. Glutamine is used as fuel by tissues such as the small intestine and **immune system**, although in addition glutamine is considered important for stimulating protein synthesis and muscle glycogen resynthesis in muscle, and ameliorating insulin resistance. From an exercise perspective, glutamine possesses a theoretical ergogenic benefit because it may:

- enhance immune function and so prevent the severity of infections or illnesses
- have an anticatabolic effect in relation to muscle protein
- promote muscle glycogen resynthesis, and
- act as an exogenous fuel source of glutamine and so spare additional glutamine being provided from muscle protein breakdown.

Intense bouts of exercise and training have often been associated with increases in susceptibility to infections. This is particularly apparent for athletes who are **overtrained**. It has been suggested that severe exercise results in a transient reduction in the body's immune function, and also that of circulating glutamine. Since glutamine is recognized as a fuel for cells of the **immune system**, it has been proposed that there is a link between the reduction in glutamine and increased risk from infections after severe exercise. At present there is limited clarity in research findings in relation to changes in glutamine following severe exercise or periods of training, and also on whether glutamine supplementation reduces the incidences of infection or improves/maintains immune function.

Glutamine provision after surgery and sepsis maintains intramuscular concentrations and also results in an increase in cell volume by improving the hydration status of skeletal muscle. An increase in **cell volume**, as observed above for creatine, is a stimulus for anabolism and increased muscle protein synthesis. Because intramuscular glutamine concentration declines in a dose-dependent manner with an increase in stress, it could be realized that there is a potential for glutamine supplementation when an athlete is likely to be engaged in strenuous exercise; exercise being the stressor. In spite of this, there is a paucity of research findings in relation to chronic glutamine supplementation and muscle **protein synthesis** in humans.

Limited findings on the effects of glutamine ingestion and **muscle glycogen resynthesis** after exercise have presented equivocal results. Glutamine infusion in animals has resulted in enhanced muscle glycogen resynthesis whilst glutamine ingestion in humans has not proved of advantage over carbohydrate feeding. Combinations of glucose and glutamine have however shown some positive effects.

Antioxidants

Many studies have shown that unaccustomed exercise, especially eccentric exercise, results in damage to muscle structure and soreness. It is recognized that **free radicals**, highly reactive chemical species, may be involved in the etiology of this damage to muscle membrane and resultant soreness. Quenching the effects of free radicals may reduce the muscle damage and pain, and thereby allow for greater training. Furthermore, free radical production is likely to be enhanced as a result of increases in oxygen consumption (which occurs during exercise) since they are produced through electron escape from the **electron transport chain**.

The body has defense mechanisms to protect itself from excessive oxidative damage. These mechanisms involve antioxidants, which are nutrients that act to prevent oxidative damage caused by free radical formation. Antioxidants may include some enzymes, such as **superoxide dismutase** and **catalase**, that help to accelerate the conversion of free radicals to less harmful agents as well as nutrients that scavenge free radicals or repair the damage caused by free radicals. **Vitamins A**, **C**, and **E**, and the trace mineral, **selenium**, are nutrients which possess antioxidant properties.

A number of studies have examined the effects of antioxidant supplementation on measures of free radical production as well as **muscle soreness** and muscle function. The evidence from selected studies using vitamin C or selenium suggests that there may be reductions in markers of free radical production and muscle damage, although the effects on muscle function are equivocal. Since training brings about responses including enhanced endogenous antioxidant activity, the requirement to supplement with antioxidants may not be necessary. On the other hand it may be prudent to consider supplementation for a short period when training increases and until the body adapts to the regime.

H1 TRAINING PRINCIPLES

Key Notes

Individuality	Individuality refers to the fact that all athletes are not the same, and that heredity significantly influences the speed and degree to which a body adapts to a training regime. Therefore a training program should take account of individual needs.
Specificity	Training adaptations are specific to the type of activity undertaken. Training for swimming is ideally performed in water rather than dry land, whilst a cyclist is better suited to cycling as a mode of training rather than running.
Progressive overload	The concepts of progression and overload are the foundation for all training. Progression means that as the training continues there is a need to increase the resistance in resistance training or produce a faster time in sets of running, swimming or cycling, whereas overload is where the muscles need to be loaded beyond which they are normally loaded.
Maintenance	Once a specific level of adaptation has taken place, this level can be maintained by the same or a reduced volume of work.
Reversibility	If an individual ceases training, the muscles (or cardiovascular system) become weaker, less aerobic or less powerful with time.
Warm-up and cool-down	Although not essentially training principles *per se*, nonetheless warm-up and cool-down should play an integral part in any training program. Warm-up implies that as a result of appropriate activity the muscle temperature is elevated from that at rest. Cool-down occurs when appropriate activity following exercise is undertaken to gradually reduce muscle temperature and aid removal of waste products from muscle.
Related topics	Bioenergetics for movement (B) Cardiovascular adaptations to Pulmonary adaptations to exercise exercise (E) (D)

Individuality

Heredity plays a significant role in determining how quickly and how much a body adapts to a training program. Other than identical twins, no two individuals have exactly the same genetic characteristics. Consequently, there can be large variations between individuals in cell growth and repair, metabolism, and regulation of processes by nerves and hormones. These individual variations may explain why some athletes can improve significantly on a certain training program whereas another may experience little or no change following the same training program. Appropriate training normally results in improvements in strength, flexibility, power, speed, aerobic power and so on, although the rates at which these changes occur vary between individuals.

Muscle fiber type is a genetically inherited characteristic, and so an athlete who has a relatively large proportion of slow oxidative (SO) fibers is unlikely to adapt to power and speed training as rapidly as an athlete who has a relatively large proportion of fast glycolytic (FG) fibers and vice versa for endurance training. Coaches and managers should be aware that lack of significant progress in a fitness parameter is not necessarily due to the athlete being lazy, but may be that the training is not wholly suitable for that person.

Another important influence in the rate of adaptation to a training program is the individual's level of fitness. In general, the greatest improvements in fitness are observed with those who are less fit at the beginning of the program. For example, $\dot{V}O_{2max}$ can be increased by as much as 50% when middle-aged sedentary males are placed on an endurance training program, yet a similar program only results in a 10–15% improvement in normal active adults. However, it should be recognized that for elite athletes a small increase in their level of fitness can make the difference between being a medallist or not even reaching the final.

It is for these reasons that individualized training programs should be employed, although for many practical considerations this is not always possible. For example athletes involved in team sports and some individual sports often train together, undertaking the same regimes. This can be seen not only in team sports such as soccer, rugby, hockey, basketball, and volleyball, but also in some individual sports such as swimming and athletics. Coaches can however target individuals to undertake extra sessions in order to redress any shortcomings.

Specificity

The concept of specificity refers not only to the mode of training and the muscle groups trained, but also to the energy systems needed to provide ATP for undertaking the activity. Training adaptations are specific to the type of activity as well as the volume and intensity of the exercise performed. A soccer player for example would not be expected to perform significant amounts of training in a swimming pool nor a swimmer train predominantly on a cycle ergometer. In these instances specificity refers to the mode of exercise.

In addition, specificity also refers to the muscle groups trained. Although most athletes would train both upper and lower body muscle groups for strength and symmetry, it is unreasonable to expect cyclists to spend a great deal of time undertaking upper body routines nor long-distance swimmers to spend too much training effort in developing their legs. The reasons are clear, in order to improve performance in their sport, athletes need not only train using a similar mode of training to their sport but also to train the muscle groups which are most likely to engender the movement. Even more specifically, athletes should be training the appropriate muscle groups in the same manner and speed likely to be used in the performance. A javelin thrower would be expected to employ training routines associated with the muscle groups and movement patterns involved in throwing the javelin.

As already mentioned, specificity also refers to the energy systems likely to be involved in generating the ATP for that sport. It seems pointless that a 100-meter sprinter would carry out 10 mile runs or a marathon runner perform many repeated sprints as a significant part of their program. The energy demands of the sport must be recognized so as to provide guidance for the types of training to be undertaken. *Table 1* highlights the energy sources for selected sports, from which appropriate training strategies can be produced.

Table 1.   Predominant energy sources for selected sports

Sport	% contribution from energy source		
	ATP-PC	Glycolytic	Aerobic
Athletics			
100 meters	98	2	0
800 meters	30	60	10
5 kilometers	10	20	70
marathon	0	5	95
Basketball	55	30	15
Hockey	20	30	50
Netball	40	30	30
Rowing	20	40	40
Rugby Union	15	25	60
Soccer	10	20	70
Swimming			
50 meters	45	50	5
100 meters	40	40	20
400 meters	20	30	50
Tennis	20	30	50
Volleyball	55	30	15

Progressive overload

Systems such as the cardiovascular system and muscles increase their capacity in response to a training overload. In order to achieve this the training program must stress the system above the level to which it is accustomed. As the body adapts to the training load there is a need to progress to a higher work level. For example, an individual who starts to leg press with 70 kg and can just about undertake three sets of eight repetitions, within a week or two should be capable of either performing three sets of 10–15 repetitions or to be using the original number of repetitions but with a resistance of 80 kg. Thus there is a progressive increase in the total amount of weight lifted (see *Fig. 1*). Likewise with sprint or aerobic training sessions, the intensity and duration of the program should increase. Progression can be achieved by increasing the distance or work

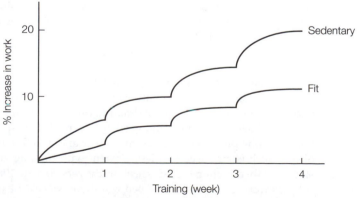

Fig. 1.   Schematic illustrating the effects of training on the % improvements in work with a sedentary individual and a relatively fit individual.

load, increasing the number of repetitions, decreasing the length of the recovery period, or decreasing the time to cover a set distance. The key to success is an increase in intensity and total training volume.

Progressive overload should be gradual. The total work done in a session should not normally exceed 10% a week. For beginners this might be easily achievable, whereas for more experienced athletes the progression is likely to be small and may take more than a week before even a 1% increase can be reached.

Maintenance

Once a set level of adaptation has resulted from a training program, this level can be maintained by the same or even a reduced volume of work. Training one day a week in order to maintain a set level of aerobic capacity is feasible as long as the intensity is maintained. However, the maintenance phase cannot last indefinitely. Training one day a week to maintain a set level of aerobic capacity is possible for a few weeks before detraining effects are observed. Therefore in the off-season, athletes are advised to undertake minimum amounts of training for maintenance. This is approximately one session a week for aerobic fitness, two sessions a week for strength, and three or four sessions a week for flexibility.

Maintenance is also an important consideration for athletes during the competitive season where it is unlikely that all the fitness parameters can be addressed during the week. In this instance coaches **periodize** training so that at particular times of a season the training emphasizes one or two fitness parameters rather than trying to improve all. Examples of this can be seen when coaches often emphasize improvements in aerobic capacity and strength through aerobic and strength training regimes in the winter months with maintenance in speed and power. Speed and power are then emphasized nearer competition when aerobic and strength training are maintained. In sports such as soccer, the pre-season training is often the time when emphasis is placed on improving aerobic capacity and strength, whilst nearer the start of the season the emphasis shifts to speed and power work. During the soccer season the emphasis is on match play and maintenance of the fitness parameters. However, if opportunities present themselves the coach may place emphasis on strength or aerobic capacity for a short periodized cycle of a week or 10 days.

Reversibility

When an athlete stops or reduces training for a significant period of time, the improvements previously observed are reversed, i.e. **detraining** occurs. Any gains in fitness through training will be lost. This reversibility of adaptations to training occurs in the muscle within days or weeks after training ceases resulting in a reduction in both maximal and sub-maximal performance. However it is not only muscle that regresses with inactivity, so does the cardiovascular system.

Individuals who have undergone a relatively short period of training exhibit the fastest rates of reversibility whereas those who have trained for long periods appear to be capable of 'holding on' to their fitness for a longer period. Aerobic fitness is generally more easily reversed than speed, probably because the metabolic processes required for power and speed involve anaerobic metabolism which is increased to a lesser extent with training and therefore lost more slowly with detraining. A classical study reported in 1968 showed that when subjects were given 3 weeks of bed rest there was a 20% decrease in $\dot{V}O_{2max}$ and maximum cardiac output. Hence, aerobic capacity can be rapidly reversed with detraining.

Warm-up and cool-down

Although not strictly training principles, warm-up and cool-down are nevertheless important components of training. Training sessions should be organized based on a warm-up, the work out, and then a cool-down. The purpose of the warm-up is to increase blood flow to the muscles and thereby deliver oxygen and metabolic nutrients for the muscles to work, and also to increase the muscle temperature so that the enzymes responsible for generation of energy can function at their optimum. In addition, it is recommended that some stretching routines are incorporated into the warm-up. Warm-up should last between 5 and 20 minutes dependent on the type of training that ensues. A low level of aerobic training demands a short warm-up whereas a vigorous power and speed session necessitates a more lengthy warm-up. The gap between the warm-up and the training should not be longer than 15 minutes or else the effects of the warm-up are likely to be lost.

The warm-up should involve large muscle groups to promote cardiac output and increase blood flow to the muscles, thereby raising body and muscle temperature. The intensity of the warm-up should not result in fatigue, and so working at 50–60% of maximum heart rate (max HR) is generally suitable. For more elite performers some short periods in the latter stages of the warm-up with a heart rate approximately 70–80% max HR is deemed suitable. Specificity of actions should also be included in a warm-up, i.e. soccer players should incorporate running sideways and backwards at speed, changing directions, jumping to head a ball, and kicking a ball.

A cool-down is desirable after a strenuous session or game in order to remove lactic acid and any other metabolites or hormones that may have accumulated, and generally to enable blood that may have pooled in the muscles to be returned to the central circulation. The length of the cool-down need only be 10–20 minutes depending on the intensity and duration of the previous training session as well as the level of fitness of the athlete. The cool-down should be performed at an intensity of approximately 70% max HR since this intensity has been shown to result in the fastest rates of lactate removal, although activity as low as 50% max HR is also desirable if the previous session has not been too intense. A stretching routine should also be used. Results from selected studies that revealed a 20 minute cool-down resulted in less muscle soreness in the following 24 hours.

H2 TRAINING FOR AEROBIC POWER

Key Notes

Interval training	Interval training results from short to moderate periods of exercise interspersed with recovery.
Long slow distance training	Long slow distance training (LSD) is performed at a relatively low intensity (i.e. 60–70% max HR) with the main objective being distance covered rather than speed.
High-intensity continuous training	This type of training is normally performed at exercise intensities approximately 80–95% max HR, with the emphasis on speed rather than distance covered.
Fartlek	Fartlek or speed play is a form of continuous training whereby the athlete changes pace during the session, i.e. slow then fast then slow than fast pacing.
Circuit training	Circuit training involves a series of selected activities performed in a given sequence. Normally associated with resistance training, circuit training has been adapted for the needs of improving aerobic power.

Related topics	Bioenergetics for movement (B)	Cardiovascular adaptations to
	Pulmonary adaptations to exercise (D)	exercise (E)

Interval training

With interval training, short to moderate periods of exercise are alternated with short to moderate periods of recovery. Research has demonstrated that athletes can undertake considerably more total work in a session if they alternate the short intense active bouts followed by recovery. The key components to interval training are the **work:rest ratio**. Depending on the energy system being trained, the work:rest ratio varies. For improvements in aerobic power and capacity, the work bout is normally from 1–3 minutes of high intensity followed by a recovery period of approximately 1–3 minutes. Interval training to enhance the aerobic system normally requires a work:rest ratio of 1:1 or 1:2. For example, an **interval set** for a swimmer could be six **repetitions** of 100 meters in around 75 s followed by a 75 s recovery between repetitions (and no longer than 120 s). After the set a longer recovery may ensue before a second set of repetitions, either of similar activity or a variation, takes place. *Table 1* provides an example of an interval training session for a swimmer. During the rest periods the athlete may either remain stationary or work at a significantly reduced pace (i.e. passive or active rest), although the heart rate should have decreased to about 60% max HR before the next repetition.

Interval training can be used for almost any sport and can be adapted by selecting the mode of training as well as the intensity of the work bout. In planning an interval session, the variables of length of the work bout, intensity of

Table 1. Example of an interval training session to improve aerobic capacity for a swimmer

Set	Repetitions	Distance (m)	Work time	Rest time
1	6	100	75 s	75 s
2	6	100	75 s	75 s
3	4	200	200 s	180 s
4	4	400	360 s	300 s

the work, duration of the rest bout, number of sets, and number of repetitions in a set need to be addressed.

Long slow distance training (LSD)

The use of LSD runs, cycles or swims involves performing exercise at a low intensity (60–70% max HR) for durations longer than the competition event. One of the beliefs for this type of training is that improvements in endurance are proportional to the volume of training. This is particularly prevalent amongst some coaches of elite swimmers with the adage that more distance in a session results in better performance. Recent evidence suggests that short, intense bouts of exercise are superior to prolonged low levels of training. Indeed, a training study reported in 1991 on swimmers demonstrated that more intense training for 90 minutes a day resulted in similar or better performances than swimmers who trained for 180 minutes a day.

However for older populations and those who exercise purely for health benefits, use of LSD may be preferable to undertaking more intense bouts of training. Under these circumstances, LSD is effective because it can be performed at a comfortable pace and, as long as the distance is not too great, a less risky way to train.

High-intensity continuous training

This type of training is performed at work intensities at approximately 80–95% max HR, and is very effective for training endurance athletes without working out maximally. Training at a constant pace which is near (but not at) race pace enhances the athlete's ability to judge pace, yet provides a very good aerobic stimulus. Serious athletes do need to train near their competition pace in order to develop limb speed, strength and local as well as cardiovascular endurance. The downside to this type of training is that it is intense and should not form the sole training strategy since overuse injuries may occur. Nevertheless it should form an important part of a performer's training since one of the key principles of training is specificity.

Fig. 1 illustrates the effect of training intensity on aerobic capacity. The figure clearly demonstrates that exercise intensities above the lactate threshold (which normally occurs between 65–80% $\dot{V}O_{2max}$) provide the best stimulus. It is likely that these training intensities may correspond to V_{OBLA}, which has been shown to significantly improve aerobic performance.

Fartlek

The Swedes have used fartlek training since the 1930s for improving aerobic capacity in distance athletes. Fartlek combines the aerobic requirements of continuous exercise with the anaerobic requirements of interspersed speed intervals, and is often undertaken in the countryside. The concept of fartlek is to run or cycle at a steady pace for a set distance and then to sprint or exercise intensely for a short distance before going back to the steady pace. This can be

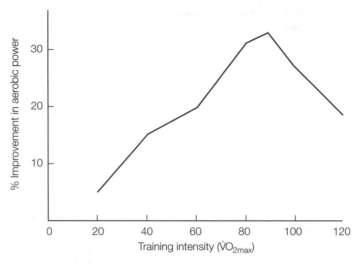

Fig. 1.   Relationship between training intensity and improvements in aerobic capacity.

achieved for example, by running steady between two or three lamp-post distances and then running briskly between one set of lamp-posts, and so on. Fartlek training provides variety and fun, yet is a good way to enhance aerobic power and capacity.

Circuit training   In circuit training a series of different activities is performed in a given sequence. A circuit normally has at least six stations wherein the individual will exercise for a given period before resting (or not) and moving on. Alternatively, each station requires the individual to complete a set amount of work before moving on. Improvements can be seen when the time taken to complete each station or the whole circuit is reduced. Although mainly used to develop strength, circuit training can be adapted for improvements in both local muscle endurance and overall cardiovascular endurance.

H3 TRAINING FOR ANAEROBIC POWER

Key Notes

Training the ATP-PC system	Training to improve the ATP-PC system involves short, very intense bouts of activity followed by recovery. Interval training is employed.
Training to improve the glycolytic system	Training the glycolytic energy system also involves interval training, although at a marginally lower exercise intensity than above and for a longer duration for each repetition.
Related topics	Bioenergetics for movement (B)

Training the ATP-PC system

The ATP-PC system is the major energy source for intense bouts of activity that last for 2–10 seconds. This type of activity is important not only for weight-lifters and field event throwers and jumpers in athletics, but also as part of many team sports such as soccer, rugby, netball, hockey, basketball, and volley-ball. A specific form of interval training is employed to improve the ATP-PC system which involves very intense bouts of high-intensity activity lasting no longer than 10 seconds with recovery periods varying between 30 and 300 seconds. Since PCr is totally replenished in around 300 seconds and 70% recovered within 30 s, the rest period can be altered either to compromise replenishment or to ensure that complete recovery is achieved. When devising the number of repetitions in a set, the fitness level of the athlete needs to be gauged.

Training to improve the glycolytic system

Glycolysis provides the significant contribution to energy when intense bouts of exercise progress beyond 10 s and are less than 60 s. Interval training in which the work bout lasts between 20 and 60 s is ideal to improve energy from this system. The rest period needs to vary dependent on whether the need is to tolerate and promote clearance of lactic acid or whether the need is to enhance the activity of glycolytic enzymes. The former can be achieved by ensuring that the rest period is between 60 and 240 s so that lactic acid levels in muscle and blood will still be elevated prior to the next repetition. The latter can be achieved when the rest period is around 20 minutes, thus allowing for clearance of muscle lactic acid in an active recovery process.

H4 TRAINING FOR STRENGTH AND POWER

<table>
<tr><td colspan="2">Key Notes</td></tr>
<tr>
<td>Fundamentals of strength training</td>
<td>It has been proposed that there are four fundamentals of strength training. These are development of joint flexibility, tendon strength, core stability, and to train muscle movement.</td>
</tr>
<tr>
<td>Resistance training</td>
<td>A program of exercise in which force is exerted against a load in order to develop strength is known as resistance training.</td>
</tr>
<tr>
<td>Power training</td>
<td>Power is the ability of muscles to generate the greatest amount of force in the shortest possible time. Although strength is a necessary component for power, strong muscles are not necessarily powerful muscles since they may not be able to generate force quickly. Power training therefore involves speed.</td>
</tr>
<tr>
<td>Related topics</td>
<td>Bioenergetics for movement (B)   Skeletal muscle contraction and control (C)</td>
</tr>
</table>

Fundamentals of strength training

In order to ensure that adaptation occurs, four fundamentals of a strength training program should be applied. These are the development of joint flexibility, tendon strength and core stability, and finally to train movement of muscles.

Most strength training exercises employ the full range of motion capable of a joint. Merely training the strength of muscles over a short range of joint motion lacks the specificity that the joint is likely to undergo during sporting actions, and may even lead to injuries if the joint motion is exceeded. Good flexibility prevents strain around these joints and stress-related injuries. Ideally, athletes should have developed flexibility whilst young.

Muscle strength improves at a faster pace than that of ligaments and tendons with the consequences that too vigorous a strength-training regime may cause injury to ligaments and tendons. The 'pace' of development for training strength may well be determined by the 'pace' at which ligaments and tendons develop strength. As with muscles, strength training leads to an increase in diameter of ligaments and tendons, thereby increasing their ability to withstand tension and tears.

In order for arm and leg muscles to develop strength properly, it is necessary for the trunk to be strong. A poorly developed trunk becomes a weak support for limbs. Strength training programs should therefore address the issue of developing core strength (i.e. strength of the trunk) before focusing on limb strength. Consequently, it is important that strengthening of abdominal and back muscles is undertaken since these often act as the shock absorbers for routines involved in jumping, lifting, throwing, and even falling. These core muscles are invariably slow-twitch fibers since they are constantly in action.

Ideally, strength training should train muscles in the movements they are likely to be used. Many skills in sports involve multi-joint actions occurring in a particular sequence. Whereas bodybuilders often train individual muscles in order to develop strength and hypertrophy, strength training for sportspersons should develop strength in the specific muscle groups.

Resistance training

Resistance training involves exercise in which the muscles exert a force against an external load. It is most commonly referred to as weight training. It is perhaps the most common method of training to improve muscle strength and enhance muscle hypertrophy. Such a training program should be individualized, be progressive, and attempt to be specific in terms of the way the muscles are likely to be used in the chosen sport. *Table 1* highlights the key points regarding resistance training for enhancement of strength. The weight should be greater than 85% of an individuals **1 repetition maximum (1 RM)**, with between 6–8 repetitions in a set. Each session should employ 3–6 sets with a recovery of about 2–5 minutes between sets. Following such guidelines ensures that progression and overload occur since the 1 RM increases with training and hence the weight lifted becomes greater as strength gains are observed. Lifting the weight relatively slowly throughout places the muscle under tension and also improves safety. Athletes need to develop strength not only in the muscle groups most associated with their chosen sport but also attempt to lift the weight in a manner similar to the actions of the muscles during a sporting action.

Table 1. Guidelines for resistance training loads

Goal	% 1 RM	Repetitions	Sets	Rest period	Speed
Strength	>85	6–8	3–6	2–5′	Slow
Power	50–80	4–10	3–6	2–6′	Fast
Muscle endurance	50–60	12–20	2–3	0.5–1′	Medium

Power training

Strength training has become an integral part of the training program for most athletes. However if the improvements in strength are unable to be transformed into sport-specific strength, then there is limited value in strength *per se*. Once gains in strength have been achieved, they must be converted to improving speed and power, where power is the ability of muscles to produce the greatest force in the shortest amount of time. The advantage of power training is that it trains the nervous system by shortening the time of motor unit recruitment (especially the fast-twitch fibers) and increasing the tolerance of the motor neurons to increased innervation frequencies.

Power training must be used to activate motor units as quickly as possible. This occurs as a consequence of adaptation in the form of better synchronization of motor units and their firing pattern, and enhanced neuromuscular coordination. It should also be recognized that there is improved co-ordination between agonist and antagonist muscles. Three types of training can be used to improve power. These are isotonic weight training, ballistic training, and plyometrics.

Attempting to move a weight as fast as possible through an entire range of motion is the standard method of power training. Free weights or weights incorporated into a resistance-training device are good means of developing power. As a general rule of thumb, relatively light weights with a fast or explosive action are needed for power training. There should be no more than ten repetitions per set and around 3–6 sets per session (see *Table 1*).

For purposes of power training, an athlete's muscle force can be exerted against various implements such as medicine balls and rubber cords. Because the muscle force normally far exceeds the resistance of these training implements, the resulting motion is explosive. The fast, ballistic application of force is possible due to quick recruitment of fast-twitch muscle fibers and effective co-ordination of agonist and antagonist muscle fibers. Examples of routines using ballistic training include various types of medicine ball throws or stretching rubber cords fixed at one end. A typical routine would incorporate 3–5 sets of 10–20 explosive repetitions with a 2–3 minute recovery between sets.

Plyometric training is a relatively more recent form of training to improve power. In essence it involves exercise in which the muscle is first loaded in an eccentric (lengthening) contraction followed immediately by a concentric (shortening) contraction. Research has demonstrated that a muscle that is first stretched before a contraction contracts more forcefully and rapidly. Plyometric training results in recruitment of most of the motor units in corresponding muscle fibers, the transformation of strength into power, and the development of the nervous system to react with maximum speed and thereby generate greater force. Examples of plyometric exercises include jumping from a low bench to the floor and immediately jumping up explosively, bounding such as in a triple jump, hopping over a low hurdle and immediately jumping up high, catching a medicine ball behind your head and immediately throwing it forwards and upwards and so on. The total number of repetitions in a training session is high – normally 5–20 sets of 5–15 repetitions with a 3–8 minute recovery.

H5 TRAINING FOR FLEXIBILITY

Key Notes

Flexibility	Flexibility can be defined as the range of motion in a joint that reflects the ability of the muscles and tendons to elongate within the limitations of that joint.
Static stretching	Static stretching is a means of improving joint flexibility by means of continuously holding a stretch in a maximal or near maximal position.
Dynamic stretching	Dynamic stretching is sometimes referred to as ballistic stretching and occurs when the joint is placed into an extreme range of motion by a fast, active contraction of agonist muscle groups. The result is that the antagonist muscles are stretched rapidly and forced to elongate.
Proprioceptive neuromuscular facilitation	Proprioceptive neuromuscular facilitation (PNF) is a stretching technique in which the muscle to be stretched is first contracted vigorously. The muscle then relaxes and is either actively stretched by contraction of the opposing muscle or passively stretched.
Related topics	Skeletal muscle contraction and control (C)

Flexibility

Flexibility is the range of motion in a joint which reflects the ability of the musculotendon structures to lengthen in the limitations of the joint. There are two types of flexibility, and these are **static flexibility** and **dynamic flexibility**. Static flexibility is concerned with the range of motion about a joint with no consideration as to how quickly it is achieved, whereas dynamic flexibility refers to the resistance to a motion in the joint. Dynamic flexibility accounts for the resistance to a stretch and is more important for athletic performances and for stiffness in joints due to arthritis.

Flexibility and stretching are important not only for athletic performance but also for undertaking everyday routines involving bending and reaching. Stretching is frequently undertaken prior to training or matches in order to prepare for the activity and possibly enhance performance of that activity as well as a means of injury prevention. In many sports, flexibility is a crucial factor in success, and this can be realized in gymnastics, diving, figure skating and hurdles. In many other team sports a degree of flexibility is needed to ensure that muscle tears do not occur when placed under stress. It should be remembered that muscles, tendons, and ligaments are the tissues most often injured. In spite of this, there is no conclusive evidence that high levels of flexibility protect against injury. In fact **hypermobility** may predispose an individual to injury.

Static stretching

Static stretching occurs when the muscle is stretched slowly and put into a position of controlled maximal or near-maximal stretch by contracting the opposing muscle group. The stretch is normally held for a period of 20–60 s. With this type of stretch the neuromuscular spindles are not activated although the Golgi tendon organs are. Consequently the result is a relaxation in the stretched muscle group and the individual may elongate the muscle even further. Because there are no uncontrolled sudden forces involved, injury is unlikely with this type of stretch.

Dynamic stretching

This form of stretching is characterized by an action–reaction bouncing motion in which the joints are placed into an extreme range of motion by fast, active contractions of agonists. As a result the antagonistic muscle groups are stretched rapidly and forced to lengthen. The quick stretch results in the myotactic reflex (a strong reflex contraction caused by the neuromuscular spindle transmitting impulses to the spinal cord) in which the rebound bounce is proportional to the original movement. Although dynamic stretches occur in most sporting activities, the use of dynamic stretching is not usually recommended as an effective means of improving flexibility.

Proprioceptive neuromuscular facilitation (PNF)

PNF is a stretching technique in which the muscle to be stretched is first contracted maximally. The muscle subsequently relaxes and is then either actively or passively stretched. The rationale for the use of PNF is that muscles relax after a contraction because the Golgi tendon organs respond to the contraction and cause the myotactic reflex which in turn relaxes the stretched muscle. The technique involves the muscle to be stretched being placed under maximal stretch by action of the agonist muscle groups using either a dynamic or static contraction. A partner then gradually passively stretches the antagonist further when it has relaxed.

I1 THERMOREGULATION

Key Notes

Overview	Exercise that takes place at a time of adverse environmental conditions presents major challenges to the human body to maintain homeostasis and health whilst meeting the metabolic demands of exercise. There are many environments that now support sporting events and the human must be well prepared to meet the physiological demands imposed. One of the most common challenges is exercise in a hot or cold climate and to understand the body's responses in these settings a basic understanding of thermoregulation is required.
Thermal balance	Physiological and biochemical processes in humans are very temperature sensitive to the point where an increase in internal body temperature of only a few degrees can lead to death. As such the human must carefully control its body (core) temperature and maintain thermal balance.
Heat gain	When considering the maintenance of thermal balance it is clear that the major process of heat gain is through internal metabolism. This is severely increased during exercise when metabolic rates are raised considerably.
Conduction, convection and radiation	Conduction, convection and radiation can be pathways of both heat loss and heat gain depending upon the gradient between the skin and the environment. Together these processes are of primary importance in losing heat from the body at rest in a thermo-neutral environment.
Evaporation	This pathway will lead to heat loss from the body via the evaporation of water from the surface of the body. This evaporation cools the surface from which it leaves. Sweating is of primary importance to heat loss during exercise.
Control of thermal balance	There are many processes that are activated in response to a hot or cold thermal challenge. The central controlling system is located in the hypothalamus in the brain. These areas receive sensory information, integrate such information and activate mechanisms for heat loss or heat gain. These responses are generally under automatic, unconscious, control. Some other processes, however, are under semiconscious control.
Related topics	Energy sources and exercise (B1) The neural system (F1) Control of energy sources (B4) The endocrine system (F2) Cardiovascular structure (E1) Exercise in hot and humid Cardiovascular function and environments (I2) control (E2) Exercise in cold environments (I3)

Overview   The addition of a range of environmental stressors to the work performed during exercise is now commonplace in many scenarios and probably represents the greatest amount of **stress** a healthy body is placed under. The combination of exercise with environments that are hot, humid, cold, at high altitude, underwater, or after significant time-zone shifts will place a significant burden on the human body as it attempts to maintain **homeostasis**. The body has natural responses to exposure to such environments and to coping with acute exercise in these circumstances. However, care must be taken to ensure the body is not taken beyond its limits of adjustment and coping to the point of injury, illness or even death. A series of **acclimatization** adjustments exists for most environments and these may protect against the deleterious effects of exercise combined with environmental stress (acclimation refers to the development of acclimatization in artificial surroundings such as a heat chamber in a laboratory). As exercise in hot or cold climates is now fairly commonplace it is important to understand the intricacies of thermoregulation and its control.

Thermal balance

Physiological and biochemical processes in the human body are temperature sensitive to the point where a small rise in temperature in specific parts of the body may lead to injury, loss of function and eventually death. To prevent this humans control very tightly internal temperature and are thus described as **homeothermic**. The maintenance of body temperature is a complicated issue when you consider the vast amount of heat produced by the body (e.g. during exercise) and also the interaction with an environment that may see significant temperature changes. It is therefore a key role of various organs (e.g. cardiovascular system) and the thermoregulatory controlling centers in the hypothalamus in the brain to maintain thermal balance.

Although temperature in different sites within the body will differ (e.g. skin surface vs. deep body) an important role for the body is to maintain body (core) temperature within the range c. 36–38°C. Maintenance of core temperature, and thus thermal balance, is achieved through a range of processes (see *Fig. 1*). Specifically these are metabolic heat production, radiation, conduction, convection and evaporation.

Heat gain

Heat gain in the body is primarily derived from all metabolically active tissues where the conversion of chemical energy to kinetic energy is only 15–25%

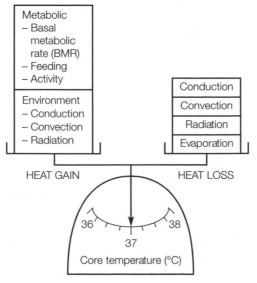

Fig. 1. Heat balance in the human body.

efficient. As energy cannot be destroyed the rest is released as heat. When exercising there is a significant increase in whole-body metabolism to meet the demands of the sporting event. The consequence of this is a significant amount of heat production. Whilst the body can store some heat it must lose most metabolically produced heat to prevent a harmful increase in core temperature and thus **hyperthermia**. The other way the body may gain heat is from the environment by processes detailed below if the temperature gradient favors heat transfer to the body from the environment.

Conduction, convection and radiation

Conduction is the transfer of heat from one object to another through direct contact. Think about what happens when you hold an ice cube in your warm hand. This process can occur between the human and various components of the environment as well as between different compartments of the body. Heat can be lost from the body in this way through contact with surfaces, clothing and equipment at a rate dependent upon the gradient between the skin and the environment. Heat loss through feces and urine flow occur via this process. It is important to remember that heat gain can occur via conduction. Think of what happens when you sit on a hot car seat for example!

Convection involves the transfer of heat between the body and a liquid or gas that moves around the warm body or body parts. As air moves over your skin it is heated to some extent before it moves away to be replaced by more air molecules that can be heated. This process can occur in reverse and warm air can heat the skin as it passes over it and again the key is the temperature gradient that will determine the direction and rate of heat transfer. Another factor that determines heat transfer is the rate of air movement around the skin. When cold air flows rapidly around the skin, heat is lost quickly and this is encapsulated in the concept of a '**wind chill factor**.'

Convection also includes the transfer of heat to, or from, water that will flow around a body if it is immersed. Because the density of liquid is greater than air, heat is lost, or gained, 20 times more quickly from someone submerged in water compared to being in air. This accounts for the very rapid loss of heat, and thus life, if people are submerged in cold water such as the seas near the polar regions.

Radiation is the loss of heat through the movement of **infrared rays** from the skin to the environment. Heat loss in this way accounts for the majority of heat loss to the environment from a human body at rest in a thermo-neutral environment. As with conduction and convection this method of heat transfer can occur from and to the body. Objects hotter than the skin will radiate a net heat gain to the skin. The sun is a significant provider of human radiant heat gain especially on sunny, clear days.

Evaporation

Unlike convection, conduction and radiation, evaporation can only result in heat transfer from the body to the environment. Because of this it is a vital route for heat loss during exercise and in hot climatic conditions. Evaporation is the process via which water on the skin is transformed into water vapor that is lost to the environment. In this process heat is lost from the surface of the skin. The loss of 1 liter of water results in the transfer of c. 2500 kJ of heat to the environment. Evaporation is primarily related to the loss of sweat released from the **eccrine sweat glands** that cover the surface of our skin. However, evaporation of water occurs from the major airways during breathing (check this by breathing on a mirror). Water loss when breathing is considered as insensible sweat loss and increases as the rate of respiration increases.

Factors that affect evaporative heat loss include an individual's level of **fitness** and/or **acclimatization** (fitter and acclimatized people sweat more and sweat earlier; see *Fig. 2*), the individual's hydration level (**de- or hypo-hydrated** people sweat less), as well as the wind speed (windier conditions promote evaporation) and the **humidity** of the environment. A **humid** environment already contains significant amounts of water vapor and thus is less able to accept water vapor from the surface of the skin. In these scenarios we often see sweat drip off the skin which is sadly providing no heat loss to the individual. It is therefore hot and humid conditions that provide the greatest threat to maintaining thermal balance when exercising and care must be taken to prevent or monitor sign of heat stress or heat stroke to the point of cancelling events if serious risk to health is perceived.

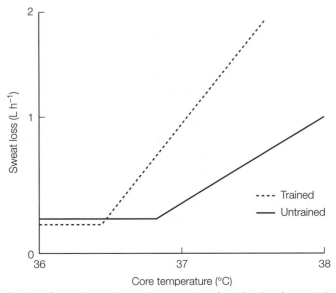

Fig. 2.   Onset of sweating and sweat rates of a trained and an untrained subject.

Control of thermal balance

It is important to understand how and where these processes are controlled (also see Sections F1 and F2). The control process begins with a myriad of temperature **sensors** or **receptors** spread around the body. These include temperature receptors in the skin that can sense heat and cold as well as receptors within the brain. All this sensory information is then sent, neurally, to the thermoregulatory centers in the **hypothalamus** in the brain. Within the thermoregulatory center there is a specific area that responds to high temperatures and a specific area that responds to low temperatures and the two areas are linked.

In essence the temperature regulation center is a thermostat that controls core temperature around 37°C. If core temperature deviates from this then a host of central and peripheral responses is initiated to promote heat gain or heat loss.

Table 1 demonstrates the processes that can be used to increase heat gain or heat loss in response to specific sensory information. The two major methods of increasing heat loss are via activating **sweat glands** and thus increasing evaporation and via increasing skin blood flow by **vasodilation** of **cutaneous** (skin)

Table 1.   Heat loss and heat gain processes

Heat loss	Heat gain
Physical processes	*Physical processes*
Cutaneous vasoconstriction	Cutaneous vasodilation
Sweating	Horripilation (Goose bumps)
	Shivering
Hormonal adjustments	*Hormonal adjustments*
Antidiuretic hormone	Catecholamines
Aldosterone	Thyroxine
Behavioural Adaptations	*Behavioural adaptations*
Remove clothing	Wear more clothes
Reduce physical activity	Increase physical activity

blood vessels. This serves to increase skin temperature and thus provide a greater temperature gradient between the skin and the environment for conductive, convective and radiative heat loss. Other strategies for reducing heat gain include behavioral decisions with conscious or semiconscious control and include removing clothing, moving to a cooler environment, eating less and drinking more.

In response to a low core temperature the body will **vasoconstrict cutaneous** blood vessels that will reroute more blood to the core as well as lowering skin temperature to provide a lower gradient for heat loss to the environment. The body will also initiate a process called **shivering**. **Shivering** increases the **metabolic** activity in skeletal muscles (without providing any external work) and thus adds to the heat gain of the body. Again other conscious or semiconscious actions include putting on more clothes, moving to a warmer environment and increasing food intake. Heat loss also promotes the release of **thyroxine** that increases metabolic rate and thus heat production.

It is important to remember that the control of core temperature is critical and small changes in core temperature will produce profound whole-body thermoregulatory responses. However, specific peripheral stimuli may still produce a local response even if internal or core temperature is not altered (for example placing an ice compress on the skin over an injury will still reduce local blood flow even if this does not affect core temperature). For the scientist studying human thermoregulation it is often important to assess both **peripheral** and **core temperatures**. The peripheral component is relatively simple as temperature thermistors can easily be placed on the skin surface. The assessment of **core temperature** is more complex. In essence we might want to know the temperature in the region of the **hypothalamus** in the brain but this is virtually impossible to achieve directly in a human subject. Other areas are more accessible, such as under the tongue, under the arm, in the ear, down the esophagus and in the rectum. However, all these measurements must be treated with some caution due to their distal location from the **hypothalamus** as well as their invasive nature.

12 EXERCISE IN HOT AND HUMID ENVIRONMENTS

Key Notes

Overview	The imposition of a significant exercise load upon a human in a hot and humid environment is possibly the most strenuous and potentially dangerous sporting scenario.
Central and peripheral blood flow	To increase all avenues for heat loss when exercising in a warm environment the body will respond by increasing skin blood flow that will require a greater heart rate and cardiac output. The skin may have to compete with the active skeletal muscle for available cardiac output.
Sweating	During exercise the primary mechanism for heat loss is via sweat evaporation. Sweating is activated at different core temperatures in all individuals and maximal sweat rates and effective heat loss are also dependent upon a myriad of factors including fitness and the relative humidity of the environment.
Heat illness	When the heat produced by the exercising muscles during exercise is not adequately removed core temperature will begin to rise and perhaps result in a range of heat illnesses from heat cramps to death.
Acclimatization	The body displays a remarkable ability to adapt to repeated exposures to exercise in hot and humid conditions. A key response is an earlier onset of sweating which promotes a more effective control of thermal balance and protects blood flow to the muscles to a larger degree. Acclimatization may take as little as 2 weeks to complete and may help offset the risk of heat illness.

Related topics	Pulmonary responses to exercise (D3)	Cardiovascular responses to
	Cardiovascular structure (E1)	training (E4)
	Cardiovascular function and	The neural system (F1)
	control (E2)	The endocrine system (F2)
	Cardiovascular responses to	Thermoregulation (I1)
	exercise (E3)	

Overview

Humans are homeothermic and are constantly exchanging (normally losing) heat to the environment. Consequently humans have to be thermogenic (i.e. they have to generate heat) normally via metabolism. However, the imposition of exercise on a human in a hot and **humid** environment is a stressful and potentially dangerous sporting setting. Exercise is especially thermogenic because active muscles generate large amount of heat as a by-product of elevated levels of metabolism. It is also important to note that unlike heat generation (an active process) the body finds it more difficult to lose heat (in essence a passive process). Exercise in a hot and **humid** environment is, therefore, replete with examples of poor performance, heat stroke and death.

With this in mind the combination of exercise and a hot and humid environment should be approached with great caution and potentially avoided if the risks are too great. Humans should be aware of the physiological responses to exercise in the heat and also be careful to note signs and symptoms of heat strain and heat illness and seek medical support if at all concerned about their own well being. Also of great importance to the exercising human is adequate preparation, which may include a process of **acclimatization** to the environmental conditions as well as sensible strategies for clothing, equipment and fluid intake.

Central and peripheral blood flow

At any given exercise intensity the imposition of a warm–hot environment (25°C and greater) or a high relative humidity (70% and greater) upon metabolism, respiratory and cardiovascular function will be marked. To attempt to maintain homeostasis in such circumstances the body increases skin temperature to promote conduction, convection and radiation as well as increasing in **sweat rate** (to be discussed in the next subsection). Skin temperature can be raised by an increase in skin blood flow and this provides a skin–environment temperature gradient to promote heat loss. This requires extra blood flow over and above the flow required by the exercising muscles. The consequence of this extra skin blood flow is an increase in **heart rate** (see *Fig. 1*). The steady increase in heart rate with steady-rate exercise intensity over time is called cardiovascular drift (see Section E) and has long been attributed to thermoregulatory mechanisms.

Whilst working at low exercise intensities there may not be a concern about this extra blood flow demand but when the intensity reaches a point at which muscle and skin are competing for portions of the **cardiac output** one of two things will happen; first performance will drop as muscle blood flow is reduced to maintain thermoregulatory function or second heat storage will increase as blood flow to the skin is compromised in an attempt to maintain performance. The latter has potential dire consequences for the health of the performer.

As well as the increased cardiovascular work, temperature has a direct effect on **respiration** and therefore **ventilation** will increase. Some evidence suggests that the active muscles use more **oxygen** at the same absolute work rate as temperature increases. Sweating is also an active process. The combined extra work of these responses will therefore increase **oxygen** uptake when exercising in the heat.

Sweating

As we have mentioned previously the process of heat loss via evaporation is the most important way of attempting to maintain thermal **homeostasis** during exercise. At rest, sweat rates are minimal and evaporative heat loss includes release of water vapor from the lungs via respiration. Even though we ventilate more during exercise it is crucial for evaporative sweat loss that we activate the **eccrine sweat glands** that cover the surface of our skin. These glands are innervated by the **sympathetic nervous system** (see Section F1) and are activated by the thermoregulatory center. Sweating is activated at different core temperatures in all individuals and is related to fitness and levels of **acclimatization**.

Sweat itself is a water-based liquid that contains some other compounds (e.g. lactate) as well as many different electrolytes (e.g. sodium). Sweat rates depend upon the level of thermal stress and the hotter the environment and the higher the exercise intensity the greater the sweat rate. Maximal sweat rates can reach 2–3 L per hour and this obviously represents a significant loss of fluid and thus

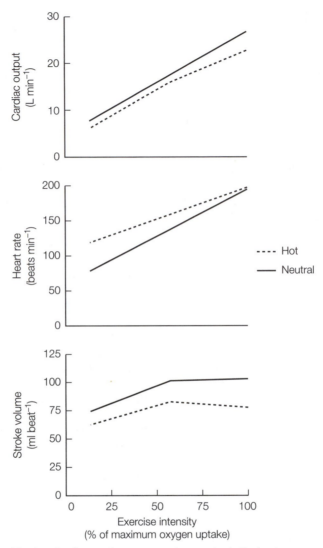

Fig. 1.   Cardiovascular responses to exercise in the heat.

body mass (see *Fig. 2*). This then leads to a decrease in **plasma volume** which has the effect of decreasing **central blood volume** and **stroke volume**. This precipitates an increase in heart rate and thus more cardiac work.

As well as the loss of fluid, excessive sweat loss can also lead to an altered electrolyte balance in internal body compartments. This stimulates the retention of sodium and fluid via the actions of hormones such as **aldosterone** and **anti-diuretic hormone**, which will result in less **urine** production.

Effective heat loss via sweating is also altered by environmental conditions. Windier weather promotes evaporation and more importantly less-humid weather also promotes evaporation. If **humidity** is high and thus water content in the air is already high, the effectiveness of sweating is reduced. Instead of

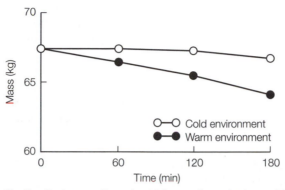

Fig. 2.   Body mass throughout 3 hours of exercise in a cold and a warm environment.

evaporating water will simply drop off the skin without removing the heat it was meant to dissipate. Thus exercise in a hot and humid environment is even more problematic in terms of performance decrement and risk of heat illness. When exercising, especially at high intensities, do not just rely on temperature as a guide to thermoregulatory strain but also consider humidity. This is contained in a measure called **wet bulb globe temperature** that assesses the combined impact of heat and humidity as well as radiant heat.

Heat illness

When humans exercise there is a huge increase in heat production at the muscle and an integrated response under the control of the thermoregulatory center to promote heat loss to maintain **core temperature**. If, however, this response is inadequate and we see core temperature rise beyond normal limits we are said to be **hyperthermic** and at risk for a range of heat illnesses.

Heat **cramps**, the least serious complaint, reflects cramping in the muscles. The exact cause of this cramping is not known but may reflect restricted blood flow as well as electrolyte disturbances associated with **hyperthermia** and fluid loss. If these occur people should cease exercise and promote heat loss in a cool environment and seek medical attention.

A more serious concern is **heat exhaustion** that is associated with fatigue, disorientation/dizziness, vomiting and fainting. Blood pressure may be low with a weak pulse. As such the body cannot cope with the thermoregulatory demands placed upon it and is likely suffering from severe fluid loss. Again people must stop exercise immediately, seek a cool environment and medical help for fluid replacement and core temperature monitoring.

Heat stroke can develop from **heat exhaustion** and is a serious medical condition that can lead to death. In these circumstances the thermoregulatory systems have failed and heat stroke is associated with confusion, loss of consciousness, lack of sweating, rapid pulse and high core temperature (>40°C). Medical attention is required urgently to decrease core temperature possibly by immersion in cold water or ice. If not treated, temperature will continue to rise and tissues such as muscles will begin to break down releasing proteins into the bloodstream (**rhabdomyolysis**) which can cause damage and failure in organs such as the heart and kidney. Death can ensue.

Heat illness can occur in all types of exercisers from those of low fitness to elite athletes. Poor preparation, insufficient fluid replacement, muscular body type and extremely high levels of motivation (over-riding warning signs) are all potential problems. Be aware that exercising conditions do not have to be hot and humid for heat illness to occur. It can be a common occurrence in polar conditions when exercising with multiple protective layers of clothing.

Prevention of heat illness includes careful planning, sensible decision-making and awareness of key symptoms. Apart from issues related to the day of exercise (is it too hot to exercise, drinking enough, etc.) one of the most important things to do to prevent heat illness is to gain some **acclimatization** to hot and/or **humid** conditions.

Acclimatization

One exposure to a hot and **humid** environment is called an acute exposure. If we expose ourselves continuously (when on holiday) or at regular intervals (via a heat chamber) the individual exposures added together will produce some adjustments to the body's physiological responses to heat and humidity called acclimatization (acclimation if exposures occur in a heat chamber). In essence, these changes occur for the purpose of making the body more efficient at losing heat and more tolerant of heat stress.

Basic changes with acclimatization (or acclimation), not surprisingly, affect blood flow and sweating. After acclimatization sweating starts earlier and so helps offset the build up of exercise-induced heat gain at an earlier point. Also sweat is more dilute, preserving electrolyte balance within the body to a greater extent. A consequence of the greater heat loss through sweating at an earlier exercise intensity or **core temperature** is then a decreased drive for blood flow distribution to the cutaneous blood vessels thus reducing **heart rate** and **cardiac output** (see *Fig. 3*).

The improved heat loss results in a reduced **core temperature** during exercise when acclimatized and thus the risk of heat illness is reduced.

The time-course for acclimatization is variable but sweating changes occur rapidly and 'full' acclimatization seems possible in 2–3 weeks. Another component of the acclimatization process that seems important is that passive heat exposure will not lead to full acclimatization and thus exercise and heat exposure must be combined.

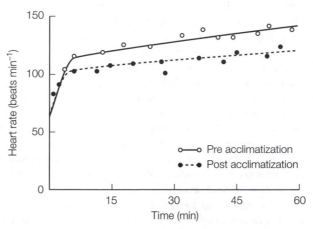

Fig. 3. The effects of acclimatization on heart rate response to exercise.

One key component of acclimatization is the earlier and higher sweat rates. The one potential negative consequence of this is an increased chance of **dehydration** and thus **hypovolemia**. Care must be taken to monitor fluid intake and recognize the increased fluid requirements after acclimatization.

13 EXERCISE IN COLD ENVIRONMENTS

Key Notes

Overview	Exercise in the cold provides a significant challenge to the individual and leads to multiple physiological adjustments. As with heat exposure, care must be taken to avoid cold injuries when exercising and acclimatization to some extent may be possible. However, we have much less research evidence to support the assertion that exercising humans can adapt to cold exposure.
Physiological responses to exercise in the cold	Peripheral circulation is reduced with cutaneous vasoconstriction and that prevents excessive heat loss to the environment. This may play a part in altered metabolism at the muscle and if muscles cool they become much weaker and fatigue sooner. A reduction in core temperature will also decrease heart rate and cardiac output.
Cold injury	If core temperature drops significantly the body is said to be hypothermic. Progression of hypothermia will lead to cardiac arrest and death. Frostbite, freezing of cutaneous tissues, is another possible consequence of cold exposure. Both hypothermia and frostbite require medical attention and are reversible if treated quickly and appropriately.
Acclimatization	There are very limited data related to the human's ability to develop some degree of acclimatization to cold exposure. This is likely to be due to the fact that simple measures to offset cold exposure, such as wearing more clothes, can be performed often with little impact upon sport or exercise performance.
Related topics	Energy sources and exercise (B1) The endocrine system (F2) Cardiovascular function and Exercise in hot and humid control (E2) environments (I2) The neural system (F1)

Overview

Exercise in the cold provides a significant challenge to the individual. Despite the fact, as noted earlier, that the body is better at heat production than it is at heat removal the stress placed on the body due to exercise in cold environments is notable. Whilst most of us may choose to exercise with more clothes or indeed exercise indoors if it gets too cold, good examples of the combination of exercise and cold exposure come from Polar expeditions and mountaineering. Acute physiological responses are apparent in cold exposure and cold injuries can occur if the athlete is not careful in monitoring signs and symptoms of injury. There is also evidence that **acclimatization** to cold will occur in certain physiological responses.

Physiological responses to exercise in the cold

Very rarely will anyone exercise in a cold environment with significant skin surface area exposed to the environment. However, whilst exercising in the cold (for example 5°C) there is a greater gradient in temperature between the skin and environment than when exercising at a thermo-neutral temperature (c. 20°C).

This facilitates heat loss and given the metabolic heat production from the skeletal muscles this could be beneficial because cutaneous vessels will not be competing with skeletal muscles for a limited **cardiac output** to promote heat loss. To protect against too much heat loss and a drop in core temperature (**hypothermia**) the body will **vasoconstrict** cutaneous blood vessels. This will serve to maintain central blood volume and core temperature. There is evidence that this will lead to a reduction in levels of circulating **free fatty acids** and thus the active muscles will rely to a much greater extent on muscle **glycogen** as an energy source.

If the environmental temperature is low enough that muscle temperature itself drops then performance will suffer due to the fact that muscle function will decrease because of altered muscle recruitment patterns and neural activity will slow down. In essence **hypothermia** will lead to an earlier onset of **fatigue**.

Cold injury

If exercise in a cold environment continues for too long, especially if the conditions are wet (immersion, snow or rain) and wind adds a '**wind-chill factor**' to the cooling process, then the body can become **hypothermic** and this can potentially lead to death. When core temperature drops below 30°C the body loses the ability to thermoregulate and a spiraling decline in core temperature is compounded by a reduction in metabolic rate. When core temperatures fall below c. 24°C then the human will likely die. It is likely that we have all heard tales of people frozen in the winter or frigid from water immersion in ice-cold lakes and a miraculous return to life when the body is slowly warmed, but these stories are rare.

One of the major problems associated with **hypothermia** is a rapid slowing of the **heart rate** due to cooling of the heart's pacemaker cells in the sino-atrial node (see Section E2). This slowing of the heart rate leaves the heart open to the triggering of fatal **arrhythmias** or the onset of tissue **hypoxia** in the heart due to low **cardiac output** and coronary artery blood flow.

If **hypothermia** is diagnosed, medical treatment is required immediately that should be directed to a slow and controlled increase in core temperature. This is often achieved with movement to a warmer environment, the addition of layers of clothing or in more serious cases the use of warm saline infusions.

Another potential health hazard associated with exercise in the cold is **frostbite**. Anyone who has any knowledge of polar trekking will no doubt be aware of the ease of contracting frostbite and the severe pain it can induce. Frostbite occurs when exposed skin freezes. Those areas likely to suffer from frostbite are exposed skin with little underlying metabolically active muscle tissue (such as the nose, the ears and the fingers and toes). The environmental temperature and exposure time required for tissue to freeze will vary but when exercising temperatures must be well below freezing point (c. −20°C). When tissue is frozen, blood flow will cease and the tissue will become hypoxic. If not treated quickly by a slow re-warming process, tissue can further break down leading to **gangrene** and eventually amputation is the only course of action to prevent further complications including death. The key is that in its early stage frostbite is reversible so medical attention and treatment must be sought quickly.

Acclimatization

One of the ways to avoid heat illness and injury is to fully **acclimatize** to the environment before engaging in intense, competitive exercise. It is therefore somewhat logical to assume that chronic exposure to cold, wet and windy environmental conditions may help in the physiological response to exercise in the cold as well as the prevention of cold-related injuries.

We are, however, still somewhat in the dark on this issue and the level of research enquiry in cold **acclimatization** does not match that associated with heat acclimatization. Again, this may reflect the pragmatism of the fact that humans rarely expose themselves to continuous cold because of the simple options of wearing clothes.

To date there is only limited evidence that prolonged and repetitive immersion in cold water will lead to an ability to maintain **cutaneous vasodilation** in the fingers for longer periods of time. This helps keep the fingers warm and active for intricate motor tasks. This area requires further research.

14 EXERCISE AT ALTITUDE

Key Notes

Overview

Exercise at altitude has fascinated humans for millennia and our understanding of its impact on performance is almost as old. The exact reasons for improved and decreased performance in different athletic events at altitude has recently been investigated thoroughly and although we do not know all the answers our understanding has considerably increased in recent decades.

The nature of altitude exposure

The key component of exposure to high altitude is hypobaria and the reduction in barometric pressure that is directly correlated to the altitude ascended. The drop in barometric pressure has a subsequent impact in reducing PO_2. A reduced PO_2 decreases oxygen delivery capacity and thus may severely limit sporting performance. Other aspects of environmental exposure include a decreasing temperature and humidity and increasing levels of wind and solar radiation.

Respiratory responses to acute altitude exposure

Because of the reduced PO_2 at altitude, the body adjusts by breathing more heavily at rest and during exercise. One of the consequences of this is a reduced PCO_2 that prompts enhanced HCO_3^- excretion via the kidneys.

Cardiovascular responses to acute altitude exposure

Adjustments to offset the relative hypoxia of altitude exposure include an increase in heart rate and cardiac output even though plasma volume and stroke volume transiently decrease. This also helps to increase oxygen delivery to the metabolically active tissues.

Metabolic responses to acute altitude exposure

A reduction in oxygen availability promotes a greater reliance on anaerobic metabolism at rest and low exercise intensities. Somewhat paradoxically lactate levels at maximal exercise are reduced at altitude.

Medical problems at altitude

These include headaches and nausea associated with acute mountain sickness. This can progress to the potentially lethal fluid accumulation associated with high-altitude pulmonary and cerebral edema. All problems can be treated with a descent to lower altitudes, supplemental oxygen and pharmacotherapy. Care should be taken to monitor for appropriate signs and symptoms especially if the ascent has been rapid.

Acclimatization

An important way to optimize performance at high altitude and potentially prevent high-altitude clinical cases is appropriate acclimatization. Alterations in cardiovascular parameters, for example an increase in circulating red blood cells, are instigated to offset oxygen delivery problems and reduce the cardiorespiratory stress of acute exposure. Despite common usage by athletes, high-altitude training does not automatically lead to improved sea level performance.

Related topics

Overview

Given the nature of human beings to explore and investigate it is not surprising that a fascination develops in many for conquering the highest peaks in the world such as Mount Everest (c. 8900 m). Humans have long been living and working at moderate altitudes (2000–4000 m) and exercise at even higher levels is now commonplace with over 100 people attaining the peak of Everest in one single day recently. It was, however, the Mexico City Olympics of 1968 (altitude 2300 m) that provoked the debate about the consequences for the exercising human of performing at altitude and this has prompted a rich vein of research projects and sports performance development in the decades since 1968.

What became apparent from situations like those in Mexico City was that some sports suffer at altitude and in others performance thrives. In essence the short-duration high-intensity sports that require the body to be moved tended to see dramatic increases in peak performance. Perhaps the classic example of this was the world record long jump of Bob Beamon in Mexico City who shattered the then world record and held it until into the 1990s. Conversely, those athletes performing for longer durations at moderate–high exercise intensities, such as the distance runners and walkers, suffered at altitude. There is a combination of factors that may affect performance at altitude compared to sea level but the primary issue relates to the barometric pressure changes associated with increasing altitude.

The nature of altitude exposure

The most important aspect of an ever-increasing altitude is that barometric (air) pressure drops consistently with elevation (see *Fig. 1*). This is due to the fact that at ever-increasing altitudes there is less air above you (and therefore less pressure on you) before you reach space.

A result of a drop in barometric pressure is a decline in the partial pressure of oxygen (PO_2). In Section D (respiratory physiology) we stated that PO_2 was determined by the fractional percent of oxygen in the air combined with the barometric pressure of the air. At altitude the percent of oxygen in the air remains at c. 20.9% but the drop in barometric pressure reduces PO_2 to the point where, at the top of Mount Everest, it is only c. 50 mmHg. If we again remember back to Section D a PO_2 of c. 50 mmHg is very close to that of mixed venous blood so this can clearly be seen as a potential problem in the delivery of oxygen to the working tissues. It is no surprise to learn that most of the people who have managed to climb to the top of Mount Everest have only done

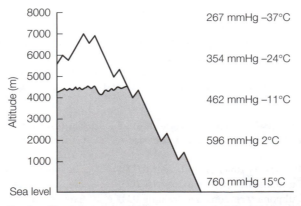

Fig. 1.   Changes in barometric pressure and air temperature with increasing altitude.

so because of the use of supplemental oxygen to breathe. Only very few have achieved this ultimate mountaineering feat with no such aids.

Of course on the plus side the reduced barometric pressure means there is less **air resistance** to movement and this has been strongly implicated in the high levels of performance in the high-intensity, short-duration sports such as sprinting.

Along with changes in air pressure there is a consistent drop in air temperature as altitude increases (see *Fig. 1*) and there is also a decrease in relative humidity that means the air is much drier at high altitude. High altitudes are also prone to periods of high winds and can receive intense solar radiation. Care must be taken in these conditions to monitor hydration status and fluid intake, to cover exposed skin to prevent frostbite and to prevent sunburn.

Respiratory responses to acute altitude exposure

Given the many changes in environmental conditions at high altitudes it is not surprising that there are huge alterations in resting and exercise physiological function in humans.

Due to the reduction in PO_2 with ever-increasing altitude there will be a corresponding decline in the gradient for oxygen transfer between the alveoli and the pulmonary circulation. This will lead to reduced oxygen saturation in the blood and a reduced oxygen transport capacity. One of the first responses to attempt to reverse this decline in arterial PO_2 is to increase ventilation and thus breathe in more oxygen over a given time period. This increase in ventilation leads to a reduced arterial PCO_2 (so-called respiratory **alkalosis**) and pH rises. To offset this, the **kidneys** excrete more HCO_3^- (an important buffer for H^+ ions associated with exercise metabolism). A consequence of the decreased PO_2 is a decreased oxygen transport, a decreased oxygen delivery to the muscles and a decreased level of endurance capacity and maximum oxygen uptake. Indeed, resting metabolic rates at rest at the top of Mount Everest get very close to the maximum oxygen uptake at such an altitude. That explains the great difficulty in performing any tasks, never mind the immense efforts of climbing.

Cardiovascular responses to acute altitude exposure

The cardiovascular system also responds to acute altitude exposure in an attempt to offset the difficulties in oxygen delivery caused by the low PO_2. One of the first things to happen is an increase in **heart rate** and thus **cardiac output** in an attempt to deliver more blood and hence more oxygen to the tissues (see *Fig. 2*). This occurs at rest and during exercise. At the same time **plasma** and **blood volume** are declining whilst **red blood cell** numbers stay the same. The result of this is an increase in hemoglobin concentration helping improve **oxygen** delivery. Conversely, the decrease in blood volume will reduce preload and thus **stroke volume** so the **heart rate** must compensate even more to increase **cardiac output**.

At maximal exercise, both **heart rate** and **stroke volume** are reduced compared to sea level and this also contributes to a decline in **maximal oxygen uptake**.

Altitude exposure produces a great increase in cardiac work, not just from the increase in heart rate but also due to the increased **viscosity** and afterload against which the heart is pumping. It is not surprising that acclimatization seeks to offset some of this cardiac load and hence can help athletic performance.

Metabolic responses to acute altitude exposure

Given the reduction in oxygen delivery with exposure to high altitude it is not surprising to see an increase in **anaerobic** metabolism at rest and during exercise (see Section B). This is clearly indicated by increased lactate production at sub-maximal exercise intensities compared to sea level. Interestingly, at

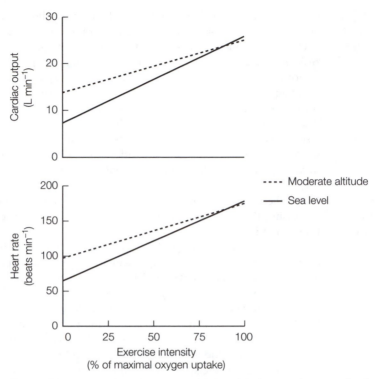

Fig. 2.   Heart rate and cardiac output at rest and during exercise at sea level and moderate altitude.

maximal exercise blood lactate production is reduced at high altitude. This **'lactate paradox'** is not fully understood but may simply be linked to the lower absolute metabolic rates attainable at such altitudes.

Medical problems at altitude

Because of the immense stress placed on the human with exposure to high altitudes it is not surprising to witness symptoms of potentially life-threatening clinical syndromes that again reflect the body's inability to maintain homeostasis.

The most common consequence of exposure to high altitude is **acute mountain sickness**. Acute mountain sickness comprises a range of symptoms, most notably severe headaches as well as sickness and insomnia. Respiratory alterations during sleep are common and may explain poor sleep patterns. Symptoms can begin quite quickly after ascent to altitude or can take a few days to develop. There is no threshold altitude for symptoms to develop but they are more severe at higher altitudes. The underlying causes of, and susceptibility to, acute mountain sickness are not fully known. Training status does not seem to prevent acute mountain sickness but a slow ascent and a sensible period of **acclimatization** seem to offset problems in most.

If acute mountain sickness is not recognized and treated rapidly and sensibly then symptoms may progress to more serious, life-threatening problems. Two problems result from the accumulation of excess fluid. Firstly, **high-altitude cerebral edema** which is fluid accumulation in the cranium can lead to confusion, loss of mental faculties and will progress to coma and death if not treated. The causes are not known but it seems to be a problem associated with ascent to higher altitudes. Treatment includes the descent to lower altitudes and breathing supplemental **oxygen**. Secondly, **high-altitude pulmonary edema** is fluid accumulation on the lungs and this will lead to **dyspnea**, shortness of breath, poor oxygenation of the blood and tissues and eventually coma and death if not treated. It occurs normally in individuals who rapidly ascend to high altitudes and should be treated in exactly the same way as high-altitude cerebral edema.

Because of the cold, dry and windy climate at high altitudes people should also monitor themselves and others for signs of **hypothermia, frostbite** and **dehydration**. Somewhat paradoxically checking for sign of **hyperthermia** may be appropriate because of the risk of **dehydration** and the significant amounts of clothing normally worn at high altitudes!

Acclimatization

As with heat and cold injuries, a sensible step in the prevention of high-altitude clinical conditions, as well as a sensible preparation for optimum performance, is the consideration of appropriate acclimatization if long periods are to be spent at high altitude. The human response to chronic altitude exposure, if performed sensibly with a slow ascent c. 200–400 m a day, includes adaptations in a range of physiological systems whose aim is to combat the reduced PO_2 in arterial blood associated with a decreasing barometric pressure. The most important changes occur in the bloodstream. The number of red blood cells will begin to increase due to the action of a hormone called **erythropoietin** released from the **kidney**. Alongside this is a fairly rapid normalization and then super-compensation of **plasma** and **blood volume**. This increases preload to improve stroke volume and reduce the **tachycardia** of initial altitude exposure. Despite the increase in plasma volume the **hematocrit** is still raised and this increases total oxygen-carrying capacity as well as carrying capacity per unit of blood (see *Fig. 3*). On the basis of improved **oxygen** carrying and delivery capacity the initial **hyperventilation** of acute exposure is reduced.

Other adaptations occur that help **oxygen** delivery to the cell including evidence of changes in the muscle itself that increases **capillary density**. However there is now growing evidence that this may be a function of muscle wasting possibly caused by reduced activity and a decrease in appetite. Associated with this is some evidence of reduced levels of key metabolic **enzymes** in the glycolytic and oxidative pathways.

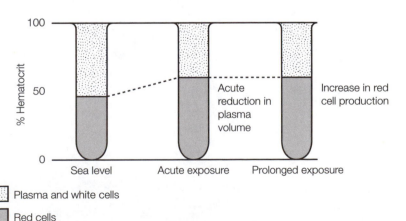

Fig. 3.   Hematocrit responses to acute and prolonged exposure to high altitude.

A couple of important points are also pertinent. The concept of 'full acclimatization' is difficult to assess but there is evidence that even prolonged periods at high altitude do not allow the athlete to perform at high altitude to the same intensity and level as at sea level. Secondly, there are a number of comparisons between high-altitude natives and subjects who have travelled for a prolonged stay at high altitude and it is clear that no matter how well trained, the visitors never seem to attain the same levels of adaptation to high altitudes as the natives.

Despite these limitations, acclimatization to the **hypoxic** conditions of altitude has become popular with endurance athletes not just as a means of improving performance at altitude but also a means of improving sea level performance. The basis of this approach is founded in the alterations in blood and red cell composition that improve oxygen transport. Theoretically this benefit should continue to exist even on return to sea level for a period of time whilst the body adjusts and down-regulates **erythropoietin** release. Interestingly, despite much anecdotal evidence the benefit of altitude training for sea-level performance is equivocal. Most people suggest this may reflect the reduced training stimulus that the subject is capable of attaining when living at high altitude. This has led to the suggestion of adopting a pattern of training at low altitudes and then living at high altitudes. This approach is difficult logistically but has prompted the boom in sales of equipment that can simulate a hypoxic environment in a sea-level laboratory (normobaric hypoxia).

15 EXERCISE UNDERWATER

Key Notes

Overview	Exercise in a hyperbaric environment, as opposed to the hypobaric nature of high altitude, does not occur on the surface of the Earth. Significant pressure is placed on the human who is active under the surface of the water, either free diving or using breathing apparatus, even with descent to relatively limited depths such as 10 m.
Responses to water immersion	The major complication of water immersion is due to the inverse relationship between pressure and volume. This means that a small descent, and therefore a rise in pressure, will have a significant effect on lung volume and gases dissolved in body fluids. The impact of immersion upon cardiovascular function is initially positive in that the hydrostatic pressure of water promotes venous return and thus an increased stroke volume and decreased heart rate for any specific exercise intensity.
SCUBA diving	Most people's diving experience will likely involve a short-term breath-hold in a swimming pool or snorkelling in the sea to very limited depths because of the requirement to take the next breath. To prolong immersion time and activity levels whilst underwater continuous breathing is allowed via the self-contained underwater breathing apparatus (SCUBA) system invented in the 1940s. SCUBA allows people to stay immersed for periods of time that primarily depends on the depth of the dive as air must be inhaled at pressures equal to the water that surrounds them.
Clinical complications of underwater activity	Owing to the immense pressure changes involved with diving and the subsequent return to sea level, allied to the use of complex equipment, the risk of injury and clinical complications with diving can be great. Care must be taken to monitor relevant signs and symptoms and respond quickly and accordingly.

Related topics	Pulmonary structure and volumes (D1)	Pulmonary responses to training (D4)
	Pulmonary function and control (D2)	Cardiovascular structure (E1)
	Pulmonary responses to exercise (D3)	Cardiovascular function and control (E2)

Overview

The opposite of the **hypobaric** effect of exercise at high altitude would be exercise in a **hyperbaric** environment at some place 1000s of meters below sea level. This sort of environment does not exist on the surface of the Earth. The Dead Sea, the lowest place on Earth, is only a few hundred meters below sea level. Even deep mines do not generally place working humans in excessively **hyperbaric** environments. The environmental conditions that can, however, place a human under great pressure, as well as requiring significant physical activity, occur when we venture underwater. Most humans may go no deeper than the bottom of the swimming pool but free diving and SCUBA supported diving has seen humans attain depths of over 100 m below the surface of the water. This places the body under immense stress.

Responses to water immersion

The most significant effect of immersion in water is the **hydrostatic** pressure placed on the body and specifically the gases in the lungs and dissolved in body fluids such as blood. The density of water is such that changes over a few meters are far more dramatic than changes over 1000s of meters in air.

As you move from sea level to just 10 m of diving, the pressure on the body has doubled and the air in the lungs has halved in volume (see *Fig. 1*). The counter to this pressure differential in long dives is to breathe air that is compressed to match the water pressure (see later section). This increases gas partial pressures.

The impact of water immersion on the cardiovascular system is to promote **venous return** and **central blood volume**. This is primarily due to the positive pressure placed on the limbs and the lower body but may also reflect **vasoconstriction** in cutaneous vessels due to the increasing cold with a descent into deep water. The result of this is an increased **stroke volume** and a decreased **heart rate** thus the cardiorespiratory response to exercise becomes more efficient (see *Fig. 2*). Care must be taken as the effects of immersion and the lowering of **heart rate** with facial immersion and cooling can lead to a **bradycardia** that may unmask **arrhythmias** in electrical conduction of the heart (see Section E2).

SCUBA diving

Early forms of diving, that still persist both recreationally and in some work environments, relied on breath holding. The limitations of this are obvious as you can only submerge for as long as the time between breaths. Whilst phenomenal free-diving records exist for depths and submerged times attained, this is not a particularly useful form of diving for those who work or simply want to enjoy the scenery underwater. The commonest method of prolonging diving times and activity underwater is via the use of self-contained underwater breathing apparatus (SCUBA). SCUBA diving apparatus was invented by the legendary diver and explorer Jacques Cousteau in the 1940s and has become popular throughout the world in work and leisure pursuits.

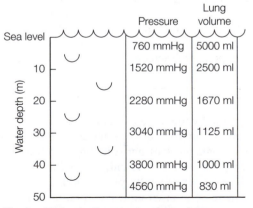

Fig. 1. Water depth and gas volume changes.

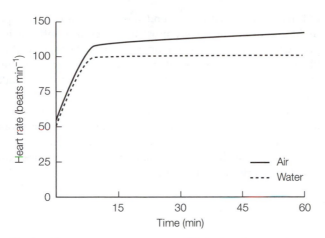

Fig. 2. Heart rate responses to exercise in air and with water immersion to the neck.

SCUBA diving involves the continuation of breathing underwater from a pressurized cylinder, normally carried on your back. This allows the diver to stay underwater for longer periods of time dependent primarily upon the depth of the dive and the size of the cylinder. The air that is breathed is pressurized to the same level as water pressure. This allows greater ventilation, greater oxygen transport and thus greater levels of activity. Expired air is simply breathed into the water as bubbles.

Clinical complications of underwater activity

Because of the great pressures, the low temperatures, reduction in light and reliance upon complicated equipment (when SCUBA diving) the potential for life-threatening clinical situations when submerged underwater is high. Care must be taken to avoid problems from a number of scenarios.

The most well known complication of exercise underwater is **decompression sickness** or the 'bends.' This occurs when a diver ascends too rapidly from a deep dive. In the descent the high partial pressure of gases inhaled will have led to greater levels of the dissolving of gases, such as nitrogen, in the bloodstream. If the ascent is too rapid the nitrogen dissolves out of solution into bubbles rather than being breathed out with a slow ascent. These bubbles may become trapped in circulation and other tissues and result in severe pain and potentially death. Bubbles in the blood (**emboli**) can disrupt circulation in key organs such as the heart and brain and result in catastrophic tissue death. The treatment will normally involve recompression, often in a land-based chamber, for the gas to dissolve back into the bloodstream and then a slow, controlled **decompression**.

The consequences of a rapid ascent can also include pressure–volume-related problems. As already noted, gas expands with ascent and if the lungs are too full with a rapid ascent they may stretch to the point of rupture of the **alveoli**. This can result in the collapse of the lung and is a serious medical condition called spontaneous **pneumothorax**. Air bubbles can also enter the circulation this way and cause similar problems to those encountered with the 'bends.' Care must be taken to adopt shallow breaths as well as to exhale and keep the mouth open during the ascent.

Other areas that may suffer from pressure-related problems are the **middle ear** and the **Eustachian tube**. Severe pain can be experienced if pressures are not equalized in ascent and descent. This can be achieved by forcibly breathing out against a closed mouth and nose and is a common procedure used in air-flight and at altitude. Care must be taken also with air in facemasks as excess pressure in the occipital region can cause damage to the eyes and associated blood vessels. Other problems include oxygen and nitrogen poisoning if increased partial pressures of gases lead to excessive build up of gases in the tissues.

Whilst such problems potentially exist with diving, extreme care must be taken in preparation for diving, including appropriate training, as well as care during the dive itself. In all instances it is recommended to dive in pairs for each to monitor the other's general health and progress throughout a dive and in the recovery time afterwards.

16 JET LAG AND EXERCISE

Key Notes

Chronobiology	The stress of competing in a hot or cold, high or low environment may be obvious to the observer. The stress of competing after a long-haul air flight (including trans-meridian travel) is less clear but can be equally debilitating. The concept of jet lag and its disruption of natural circadian rhythms are receiving much research attention and some guidelines for reducing the impact are being formulated.
Circadian rhythms	A circadian rhythm is a repetitive and daily rhythm is some aspect of human biology. This could be in physiological parameters, such as core temperature as well as biochemical and/or psychological processes. There is clear evidence of a subsequent circadian rhythm in human performance of different types.
Jet lag	Of major concern to some athletes as well as recreational travellers is the disruption to circadian rhythms that occur with long-haul flights when multiple time zones have been crossed. The consequence of such travel and disruption to circadian rhythms is a feeling of fatigue and lack of alertness, generally referred to as jet lag.
Prevention and treatment	The prevention of symptoms of jet lag can involve decisions and actions made prior to travel, during travel and after arriving at your destination. A number of preventative or treatment regimes have met with mixed degrees of success.

Related topics	Cardiovascular function and control (E2)	Thermoregulation (I1)
	The neural system (F1)	Exercise in hot and humid environments (I2)
	The endocrine system (F2)	

Chronobiology   Whilst the environmental stresses of exercise in hot, cold, **hypobaric** and/or **hyperbaric** environments are quite obvious the stress of competing in a distant geographic location (even if the temperature and altitude are similar to home) after rapid air transit can be considerable. These stresses are somewhat less well understood than thermal and barometric challenges and are open to great individual variability. Even so the impact on the performance of the athlete may be considerable and as such these problems should be understood and offset if at all possible.

Chronobiology is the study of the natural rhythms that occur within humans. Evidence for cyclical rhythms in many aspects of human physiology, biochemistry, psychology and performance is apparent. The simplest example is the daily (circadian) variation in core body temperature. Other more complex rhythms may take longer such as the female **menstrual cycle** over a month.

Rhythms are such that they are repetitive, over different timescales. However, many biological rhythms can be affected, or disrupted, by individual factors such as fitness, activity levels, sleep loss, shift work, drug use and the consequences of trans-meridian air travel (jet lag). Given the rise in recent decades of both international sporting events and simple, cheap air travel the occurrence, consequences and understanding of jet lag has increased markedly. Currently there are a host of ameliorative measures that athletes can adopt to try to reduce or remove the symptoms and consequences of jet lag.

Circadian rhythms

A circadian rhythm is a repetitive and daily rhythm in some aspect of human biology. The classic example of this is the circadian rhythm in **core temperature** (see *Fig. 1*). This rhythm repeats every 24 hours and shows a peak in the early evening and a nadir (low) in the early hours of the morning. The rhythm in core temperature is quite robust and exists even if environmental conditions change or exercise is imposed.

The circadian rhythm in core temperature is considered to be very important and possibly a controlling rhythm for many other aspects of human physiology, biochemistry, psychology and performance that have similar patterns of change over the day.

The control of biological rhythms is a complex area. Evidence suggests internal (endogenous) and external (exogenous) controlling factors. The link between various circadian rhythms and external zeitgebers 'time-givers' such as the **sleep–wake** cycle, the **dark–light** cycle and social interaction suggest an important role for exogenous controls. There is some evidence that an internal body clock is also important in controlling rhythms. Although the exact location of the internal body clock is still debated some evidence points to the **suprachiasmatic nuclei** in the **hypothalamus** and the **pineal gland**.

Jet lag

The key issue with most circadian rhythms is their relative stability over repetitive daily cycles. However, they can be disrupted by alterations in factors such as the **sleep–wake cycle** and the **light–dark cycle**. Disruption to rhythms often results in feelings of lethargy and fatigue. For example, the loss of sleep in workers who change from day to night shifts can reduce physical and mental parameters and has been associated with work-related errors.

Of major concern to athletes as well as recreational travellers is the disruption to circadian rhythms that occur with long-haul flights over many **time zones**. For example a flight from London to Chicago may take 8 hours. When the

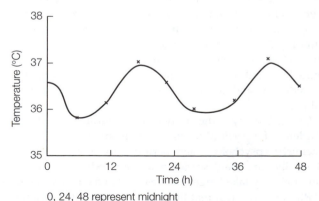

0, 24, 48 represent midnight

Fig. 1.   Circadian variation in core temperature.

aircraft takes off it might be 9 pm. However given the 6-hour **time zone** differ-ence by the time of the landing it will be 11 pm locally. This will have placed the 24-hour rhythm for core temperature, stable before the flight, in an asyn-chronous pattern with the **light–dark** cycle at the destination in Chicago. The added consequences of a loss of sleep and disruption in **sleep–wake** periods will also ensue. The consequence of such travel and disruption to circadian rhythms is a feeling of fatigue and lack of mental alertness potentially associ-ated with other changes in sleep patterns, appetite and gastrointestinal activity. This is generally referred to as jet lag. The underlying physiological conse-quences of such travel occur in all humans, however it is difficult to predict in whom symptoms will be seen and in whom the effects may well impinge upon performance. The time to readjustment or resynchronization of circadian rhythms with the **light–dark** cycle will depend on the length of travel and the number of time zones crossed. A rough guide is an adjustment period of 1 day for every time zone crossed. Interestingly, east–west travel seems to present fewer jet lag problems and a quicker adjustment to the local environment. This may be because the initial lengthening of the day may be closer to the natural, free-running biological rhythms in the body that are constrained by the lunar cycle.

Prevention and treatment

To avoid the decline in sports performance associated with jet lag, athletes are often guided in a number of directions in an attempt to remove or reduce the symptoms.

Advice pre-travel probably starts with when to travel and this should be set to allow appropriate time to **acclimatize** prior to competition. Often there are limitations in choosing the time, duration or route of the flight and so pre-flight preparation choice may be limited. Jet lag may be exacerbated by differences in environmental conditions between the points of departure and arrival. Therefore, it is important to consider preparations for different environmental conditions in the pre-flight preparation.

Once travelling an important task is to assume the time of the destination as soon as you board the flight. Changing your watch helps. The important aspect of this change is to adopt activities at the destination that are appropriate. If it is midday at destination when your flight departs then try to avoid sleep, etc. Adopting destination **sleep–wake cycles** will help adjust at the destination even if sleeping is difficult on board or you are very tired. Also consider a couple of other points for the flight. You will be cramped and seated for most of the time. Try to remain active and take frequent walks around the cabin (this may be useful in preventing deep vein thrombosis that has received a lot of recent media attention). The cabin will be warm (but usually comfortable) and very dry. **Dehydration** and dry skin are common problems so avoid alcohol and tea/coffee and instead drink water or fruit juices.

The adoption of destination time during the travel will help on arrival and should be continued. Even if very tired in the middle of the day avoid a long sleep and attempt a short nap, possibly with a shower to help wake you up. Correspondingly adopt the timing of food and training at the local destination as soon as you get there. This also relates to exposure to light. Adoption of the **light–dark cycle** of the destination is also important. A common problem of the flight and arrival times are feeling tired when you should be alert and alert when you should be tired as you adjust your circadian rhythms. Some have suggested the use of either sleeping pills at the new night-time or stimulants (such as caffeine) during periods when activity and alertness are required. Conclusive research evidence is not available to support these suggestions and care must be taken not to infringe upon doping rules in sports people.

Another intervention to receive some research attention has been the ingestion of **melatonin,** once you have arrived at the travel destination. **Melatonin** is normally secreted from the pineal gland during the night hours. Evidence from the few studies that have investigated oral **melatonin** supplementation during the evening when at the destination has reported some reduction in jet lag symptoms. However, because of the lack of clinical trials, official guidance to athletes is more prudent suggesting the use of **melatonin** only if it has been used before with no side effects.

J1 ENERGY BALANCE

Key Notes

Energy balance	The maintenance of body mass and body composition over time requires that energy intake is equal to energy expended. When this is achieved, the individual is in energy balance. Factors such as dietary habits, environmental factors, and genetic makeup influence energy balance.
Macronutrient balance	Dietary factors which are implicit in the energy balance equation include the ingestion and oxidation of macronutrients. Therefore nutrient balance occurs when there is a balance between the intakes of carbohydrates, fats, protein, and alcohol, and their rates of oxidation.
Energy expenditure	Energy expenditure is one half of the energy balance equation. Components of energy expenditure include resting metabolic rate, thermic effect of food, and thermic effect of activity.
Energy intake	Energy intake is one side of the energy balance equation and is concerned with the amount of energy ingested through food and drink. Collection and assessment of dietary intake are the most frequently used procedures for determining energy and nutrient intakes.

Related topics	Estimation and measurement of energy expenditure (A4)	Fluids (G3)
	Energy sources and exercise (B1)	Training for aerobic power (H2)
	Energy stores (B3)	Physiological benefits of exercise (K2)
	Energy for various exercise intensities (B5)	

Energy balance

Individuals whose body mass and body composition remain stable over a period of time are in **energy balance**. This occurs when the energy expended over the period of time is matched by the energy consumed. The human body mass remains relatively stable over a considerable period of time in spite of increases or reductions in food intake. For example, a researcher ingested an average of 7.42 MJ day^{-1} for one year then increased the food intake to 9.24 MJ day^{-1} for the second year, and finally an intake of 10.1 MJ day^{-1} for the third year. In spite of a 25% increase in daily energy intake from the first to the second year, and a further increase of 9% in the third year, there was only a small increase in body mass during this period. Since there was no change in lifestyle in terms of activity, the body had adapted to the higher energy intake by increasing metabolism. If the latter had not occurred there would have been a substantial increase in body mass over this period.

Some individuals are capable of increasing body mass relatively quickly and easily by consuming small amounts of extra food whereas others find it difficult to put on weight even if they consume larger amounts. Inherited genetic factors relating to the ability of the body's metabolism probably result in these differences. Differences in **metabolic rate** between individuals may be due to genetic factors, gender differences, and age-related factors.

Macronutrient balance

Changes in the type and amount of the macronutrients, carbohydrate, lipid, and protein, and their **rates of oxidation** are an integral component of energy balance. In general, increases in the intake of carbohydrates and protein lead to increases in their rate of oxidation, whereas an increase in fat intake fails to increase fat oxidation. Therefore increases in dietary fat intake are likely to lead to an increase in fat deposition whereas increases in the intake of the other **macronutrients** does not follow a similar pattern.

A study on three lean men who were overfed and also underfed for 12 days highlighted the precise regulation of carbohydrate and fat balance. In the overfed condition, the men consumed 33% more energy than expended. The diet provided 50% energy from carbohydrate, 35% from fats, and 15% from protein. Carbohydrate and protein oxidation matched intake, whereas fat oxidation was not sensitive to intake. In the latter case, the men oxidized only 60 g of the 150 g day^{-1} ingested. The net effect was an increase of 3 kg of weight over the 12 days. In the underfed condition, where energy intake was 67% less than expended, the carbohydrate and protein oxidation again matched the intake whereas fat did not. The subjects ingested 20 g fat yet oxidized 60 g day^{-1} and consequently lost 3 kg. *Fig. 1* illustrates the relationship between macronutrient intakes and their rates of oxidation.

When carbohydrates are ingested there is a stimulation of glycogen storage and carbohydrate oxidation, and an inhibition of lipid oxidation. The amount of carbohydrate not stored as glycogen is considered to be oxidized in almost equal balance to that consumed. In normal-weight individuals conversion of excess carbohydrate to fat does not occur readily. It is only when large amounts of carbohydrate are consumed over a number of days and when energy intake exceeds expenditure is there conversion of carbohydrate to fat.

In a similar fashion the body adjusts to a range of protein intakes by altering the oxidation rate of dietary protein. Nitrogen balance occurs when the protein intake matches protein breakdown. Once the body has met all **anabolic** needs from protein, any excess amino acids are converted to glucose with the elimination of the amino group as urea. The glucose can then be used in oxidation.

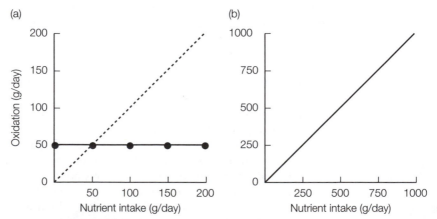

Fig. 1. Rates of oxidation compared with nutrient intake for fat (●—●) and protein (-----) in (a), and carbohydrate (——) in (b).

Inadequate intake of either energy or carbohydrate results in greater protein breakdown whilst excess intake of energy or carbohydrate spares muscle protein breakdown. Excess protein intake may spare use of fats as an energy source and so contribute to fat storage from lipids ingested rather than amino acids being converted to fats.

Fat balance is not as precisely regulated as protein and carbohydrate balance. Because most research shows that excess fat intake does not lead to an increase in fat oxidation, it appears that excess energy eaten in the form of fats is stored as triglycerides in adipose tissue. Over the long term, a positive fat balance due to excess energy intake leads to a gradual increase in body fat stores. Eventually, due to increased levels of circulating fatty acids the body will start to oxidize more fats and so a new fat balance is reached (see *Fig. 2*).

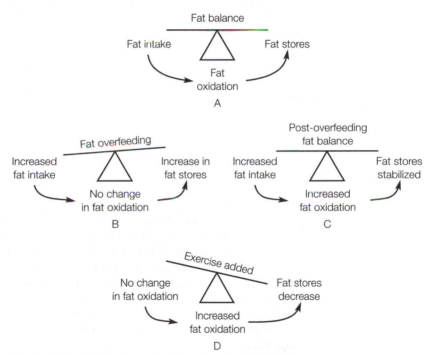

Fig. 2.   *Schematic to show adaptation to excess fat intake.*

Energy expenditure

Energy expenditure is one side of the energy balance equation, and any changes in energy expenditure without changes in intake can result in weight loss or weight gain. Daily energy expenditure is made up from three major components. These are resting metabolic rate (RMR), the thermic effect of food (TEF), and the thermic effect of activity (TEA). The units of measure of energy expenditure are the **kilojoule (kJ)** or the **kilocalorie (kcal)**, where a kJ is the energy used to move a 1 kg mass 1 meter by 1 Newton and a kcal (or Calorie = 1000 calories) is the amount of energy needed to increase the heat of 1 kg of water by 1°C, and 1 kilocalorie is equal to 4.2 kilojoules (i.e. 1 kcal = 4.2 kJ).

The **resting metabolic rate (RMR)** is the energy necessary to maintain the various systems of the body and to regulate body temperature in the resting state. To measure RMR a person must be lying down in a quiet and comfortably warm room, they must have fasted for around 12 hours or more to ensure no energy is being expended in digestion, absorption and assimilation, and must be free from stress and illness, and not on medication. For most sedentary individuals the RMR accounts for around 60–80% of the total daily energy requirements, whereas for active individuals the RMR can account for about 30–45% of total daily energy expended. A number of factors affect RMR, and these include age, body mass, sex differences, genetics, environment conditions, dieting, exercise, and illness.

The **thermic effect of food** (TEF) represents the increase in energy expended above the RMR which is due to the ingestion of food throughout the day. The TEF includes the energy cost of digestion, absorption, transport, metabolism, and storage of the food consumed, and accounts for approximately 6–10% of the total daily energy expenditure.

The **thermic effect of activity** (TEA) includes the energy cost of all daily activities above the RMR and TEF. It is the most variable component of energy expenditure because it involves all voluntary activities such as sitting, walking, cooking, ironing, and washing, as well as exercise. The TEA may only be 10–15% of the total daily energy expenditure in sedentary individuals or as high as 50% in physically active persons.

Energy expenditure can be assessed by direct or indirect calorimetry. **Direct calorimetry** measures the amount of heat generated by the body whereas **indirect calorimetry** estimates the heat produced by the body by measuring the amount of oxygen consumption and carbon dioxide production. Direct calorimetry demands the use of a closed, air-tight chamber in which the heat produced by the person in the chamber warms water surrounding the chamber. This method is expensive since it employs sensitive and sophisticated equipment (see *Fig. 3*).

Indirect calorimetry, because it measures respiratory gases, either uses a respiratory chamber or ventilated hood or mouthpiece and gas analysis system or mouthpiece and Douglas bag collection system. Energy is calculated from the amount of oxygen consumed. Taking in 1 liter of oxygen represents approximately 4.8 kcal of energy being expended if a mixture of carbohydrates and fats is oxidized. If the individual has been eating high carbohydrate meals, 1 liter of oxygen represents 4.9–5.0 kcal, and when on a low carbohydrate diet the energy from 1 liter of oxygen consumed is about 4.7 kcal. The slight variations in energy expenditure depend on the type of diet and are due to the **respiratory exchange ratio (RER)**:

$$RER = \frac{\text{volume of carbon dioxide produced}}{\text{volume of oxygen consumed}}$$

The ratio varies dependent on the fuel being used, since the oxidation of pure carbohydrates results in an RER of 1.0 whereas pure fats produce a value of 0.7. The term RER refers to whole body oxidation whereas a synonymous term, the **respiratory quotient (RQ)**, refers to cellular oxidation.

To determine the energy expended at rest using indirect calorimetry, an individual would lie on a bed with a ventilated hood (a Perspex canopy in effect) placed over their head and neck or would breath into a mouthpiece from which the expired air would be analyzed. Using this method, the oxygen consumption and carbon dioxide production would be determined over a given period of time (normally 20–30 minutes). The energy expended would be calculated from these data using well-established equations. In addition, the urine nitrogen content should be determined to account for the likely protein use.

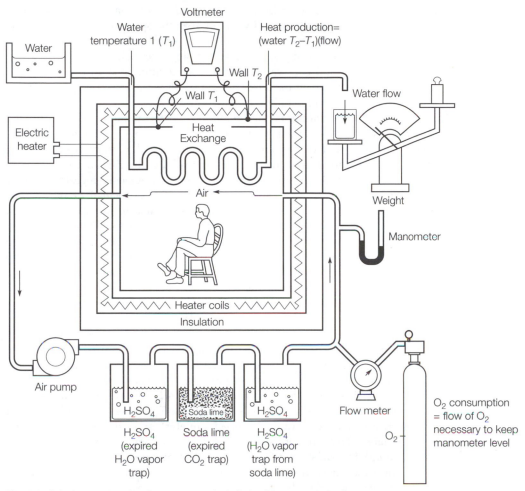

Fig. 3. Calorimetry chamber for measure of whole-body energy expenditure.

To determine the energy expended carrying out various sporting activities using indirect calorimetry, it would also be desirable to measure respiratory gases. Although there are now available gas analyzers which are light (around 1–2 kg) and portable, and so can be used to measure respiratory gases in sports such as tennis or athletic events, many sports are too invasive to use gas analysis for measures of energy expenditure. In these circumstances it is necessary to employ a **heart rate/$\dot{V}O_2$ relationship**.

The HR/$\dot{V}O_2$ relationship is normally a linear relationship which is established for each individual in a laboratory setting. The individual would walk, walk briskly, jog, cruise, and run fast on a treadmill for 4 minutes at each pace. During these bouts on the treadmill, the HR and the $\dot{V}O_2$ are measured in the last minute, and a graph of HR v $\dot{V}O_2$ produced (see *Fig. 4*). The individual can then have their HR monitored via telemetry during their chosen sport. The mean heart rate data are converted to mean $\dot{V}O_2$, and then the energy expended calculated by multiplying the $\dot{V}O_2$ in liters by 4.82 to convert to kcal. Clearly assumptions are made using this method, in particular that there is a mixed carbohydrate, fat and protein oxidation providing the energy with no account being taken for diet or exercise intensity. Furthermore, factors such as stress and environmental conditions will vary between the sport being played and the laboratory setting, and that sports involve movements that are not linear. The game of soccer, for example, involves passing the ball, jumping, tackling, running sideways and backwards, and of course running with the ball. These activities are not normally included in the HR/VO$_2$ relationship and invariably lead to an underestimation of energy expenditure.

More recently the **doubly labeled water (DLW)** technique has been validated as a measure for determining energy expenditure over a prolonged and free-living period. The DLW is an indirect form of calorimetry based on the differences in elimination of the **stable isotopes** ^{2}H (deuterium or heavy hydrogen) and ^{18}O (a stable isotope of oxygen) from a previously ingested dose of water containing these two stable isotopes (i.e. ^{2}H$_2$^{18}O instead of the normal ^{1}H$_2$^{16}O or simply H$_2$O). The deuterium is only eliminated in water excreted from the body (^{2}H$_2$O), whereas ^{18}O is eliminated both as water (H$_2$^{18}O) and in carbon dioxide (C^{18}O$_2$). The difference in rates of elimination between the two is due to carbon dioxide production, which in turn reflects the amount of energy expended. The method of DLW is non-invasive since the only measures made are on urine collected at baseline and then after a minimum period of about 3 days to a maximum of 3–4 weeks. A **mass spectrometer** is used to detect the levels of stable isotope in the urine water. DLW has now become the 'gold standard' method of assessing energy expenditure in a free-living environment. Of course because of the length of time, the method cannot be used for determining energy expenditure in single bouts of activity or individual sports. Although DLW is a valid and reliable measure, it is expensive.

In the last 15 years various **motion analyzers** have been investigated as a means of assessing daily energy expenditure. These systems either detect motion in a single plane (**uniaxial accelerometer**) or in three planes (**triaxial accelerometer**). They are very small and light accelerometers, which are usually worn around the waist or a limb, and detect amount of motion over a set period of time (normally a day). Poor correlations between the uniaxial accelerometer and DLW have been found, whilst the correlations with the triaxial accelerometers have been good.

Other methods of assessing energy expenditure include the use of activity diaries or logs, which have to be filled in by the individual, and also predicting energy expenditure from equations. The latter is used extensively and is based upon the fact that the total daily energy expenditure comprises of RMR, TEF,

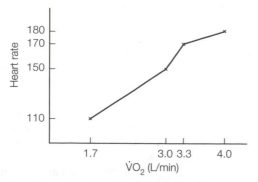

Fig. 4.   Graph illustrating the heart rate/V̇O$_2$ relationship.

and TEA. RMR can be estimated from one of a variety of equations based on age, sex, and level of activity of the individual. The next stage is to multiply the value of RMR with an activity factor, which is normally 1.3–1.6 for sedentary, 1.0–1.2 for bedridden or 1.9–2.4 for very active persons.

Energy intake

Energy intake can be estimated from a variety of techniques which in effect determine the amount and type of foods consumed over a period of time. Such techniques include **weighed food intake** and **dietary recall**.

To undertake a weighed food intake, the individual is required to weigh and write down everything they eat and drink over a period of time. Normally, if a meaningful average of macronutrient intake is to be collated and analyzed, at least 3 days and preferably 7 days are needed. The days need not be consecutive, although they should include the range of days in a week. Weekday and weekend findings are necessary. Once completed, the analysis of the recorded food is undertaken using a computerized nutrient analysis program. These programs enable determination of the energy content of all the foods ingested, and so provide a daily (or weekly) total not only of the energy content but also a breakdown of the amounts of individual macronutrients in the food. The reason for analysis of foods ingested over 3–7 days is so that a meaningful daily average can be calculated. If the data were only assessed over 1 day, an atypical day may have been selected and so introduced bias.

Dietary recall involves the individual recalling the previous 24-hour food intake with a trained dietician. The dietician will be able to assess the portion sizes that have been consumed and so arrive at approximate weights of food eaten. Analysis of the foods still requires the use of a nutrient analysis program.

J2 BODY COMPOSITION

Key Notes

Body composition, health and sport	The human body is composed from many major components at the cellular and tissue levels. These include water, minerals, protein, and fat. Increases in the levels of the fat component are detrimental to health and also sports performance. On the other hand, increases in the protein component result from more muscle mass and hence are beneficial to athletes. The mineral component is mainly associated with bone. The density of bone can be problematic in the elderly where osteoporosis arises.
Densitometry	Densitometry is based on the theory that the proportions of fat mass and **fat-free mass (FFM)** can be calculated from the densities of the known body compartments and the whole body.
Skinfold measures	Since subcutaneous fat is the largest depot of body fat, measurement of these fat depots may be useful in estimating the total body fat.
Body mass index (BMI)	Various height and body mass charts exist as a means to assess whether an individual has low, normal or high body fat. The BMI is an index of height and body mass which has been used to gauge the level of obesity of an individual.
Dual-energy X-ray absorptiometry (DEXA)	DEXA is a technique which has been employed to determine the bone mineral density of individuals, but is also able to determine the fat and fat-free components of regions of the body and the whole body.
Bioelectrical impedance	Bioelectrical impedance determines the resistance to the flow of an electrical current in the body. Normally the electrodes are placed on one ankle and one wrist.
Near infrared reactance (NIR)	This technique uses a spectrophotometer and a fiber-optic probe which emits a beam of electromagnetic radiation in the near infrared region. The beam is directed at a site such as the biceps, and the returned, reflected energy is detected by the spectrophotometer.
Related topics	Exercise, fitness and health (K1) Screening and exercise testing (L1)

Body composition, health and sport

The assessment of body composition is not only common in sport and exercise sciences but also in medicine. Most of the interest is in quantifying body fat in relation to health and to sports performance. Consequently, a number of techniques for assessing body composition have been developed over the years. The human body is made up of about 50 elements at the atomic level, of which 98% is due to combinations of carbon, oxygen, nitrogen, hydrogen, calcium, and phosphorus. At the molecular level, the body is essentially made up of the chemical compounds water, carbohydrates, protein, fats, and minerals, and these are organized into cells which are the basis of tissues and organs. The human tissues comprise of adipose tissue, muscle, bone, nerves, and epithelial tissue. The relative amounts of adipose tissue, muscle, and bone are of importance for health and sports performance, and as a result of methods of body composition analysis have been categorized into **fat mass** and **fat-free mass (FFM)** (also referred to as **lean body tissue**).

Fat mass includes adipose tissue, whereas FFM includes water, protein, and minerals. Large amounts of fat mass are associated with health problems. In essence a large fat mass results in obesity and the various health problems associated with being obese (e.g. cardiovascular diseases, diabetes, cancers, etc.). A large fat mass is also inappropriate for a sports performer, where there is a requirement for low body fat and increased muscle mass. There is an inverse relationship between fat mass and performance of activities that involve jumping or running, although not for distance swimming. A high **% body fat** results in impaired **aerobic capacity** and also high intense bouts of exercise where the need is for a high force-producing mass. Muscles are a force-producing mass but adipose tissue is not.

There is substantial evidence for FFM being positively related to performance of activities that require application of force such as in weight-lifting and throwing, although too much muscle bulk may impede jumping and running. Various sports demand variations in the amount of FFM and fat mass. In the case of the latter, the normal means is to report % body fat. The normal range for % body fat is 10–20% for young males and 20–30% for young females. Levels above the upper range constitute obesity, whilst for sports performers the % body fat scores are expected to be in the lower half of the range, i.e. 10–15% and 15–20% for males and females, respectively. Hence it is important to be able to measure FFM and fat mass.

Densitometry

Densitometry assesses the density of the body and is usually undertaken by **underwater weighing**. This has been recognized as a 'gold standard' in body fat determination. The method is based on weighing an individual in air and then determining their volume by immersion under water, having expelled out as much air as possible from their lungs. A body which has a high amount of fat displaces more water than a body with a great amount of FFM. The equation is based on Archimedes principle:

$$\text{density} = \frac{\text{mass}}{\text{volume}}$$

and so enables the whole body density to be calculated, from which % body fat may be determined from standard equations.

Skinfold measures

Since subcutaneous fat is the largest depot of body fat, measurement of these fat depots may be useful in estimating the total **body fat**. However, caution should be undertaken with this form of assessment since patterns of subcutaneous fat depots vary, as do the proportions of fat in various fat storage areas. These considerations have resulted in the numerous equations and sites from which % body fat can be estimated from skinfold measures. The skinfold method measures a double fold of skin and subcutaneous fat by means of calipers which should apply a constant pressure over the measurement site (see *Fig. 1*). Two major assumptions are made in employing skinfold measures. First, that the limited number of measurement sites is typical of any other

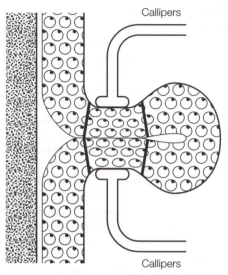

Fig. 1.   Illustration of a section through a skinfold site.

subcutaneous sites not measured, and second that the selected sites are repre-
sentative of fat depots in non-subcutaneous sites (i.e. body viscera, interstitial
spaces, bone marrow, etc.). The first assumption is not strictly held since
patterns of subcutaneous fat distribution vary across individuals. As regards the
second assumption, there is some evidence of a positive relationship between
the amount of subcutaneous fat and non-subcutaneous fat, although this relationship changes with age.

The most frequently used skinfold sites for estimating % body fat are the four
sites of biceps, triceps, subscapular, and suprailiac (*Fig. 2*). Durnin and
Wormersley (1974) used these four sites on 481 men and women and came up
with the following predictive equation:

$$D = 1.1631 - 0.0632 \times \log_w (\Sigma 4 \text{ SF})$$

After calculating density (D), the equation of Siri can be used to calculate %
body fat:

$$\% \text{ body fat} = \frac{495}{D} - 450$$

An alternative method of using skinfold measures is to sum the scores from the
sites and use the total as an index. This method does not result in a % body fat
estimate but allows values to be checked against norms for the individual or
from a cohort of players in a team.

Body mass index (BMI)

Height and body mass charts have been used for many years as an index of
evaluating relative body mass. They are easy to use and do not require expensive equipment, making them easy field measures. It would appear therefore,
that weight per unit of height could be a convenient and useful expression
which reflects body composition. The BMI is just such an index of height and
weight, and has been used extensively. The equation for BMI is:

$$\text{BMI} = \frac{\text{body mass (kg)}}{\text{height}^2 \text{ (m)}}$$

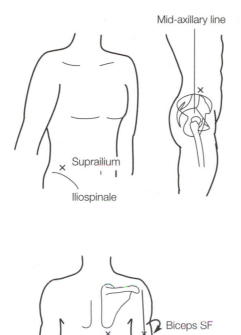

Fig. 2. Four skinfold sites. (a) Biceps, (b) triceps, (c) subscapular, and (d) suprailiac.

Overweight is defined, in BMI terms, as being between 25–29.9, and obese as above 30. The BMI score provides no information about % body fat, rather it is an indicator of adiposity. No account is taken of gender or age, and some anomalies do happen. For example, a bodybuilder of short stature with a dense muscle frame could have a high BMI yet have a very low actual % body fat. Nonetheless, BMI is generally a useful measure of obesity as long as the 'extra' weight is due to fat and not muscle. For some athletic populations the BMI might prove problematic.

Dual-energy X-ray absorptiometry (DEXA)

DEXA involves passing a dual-energy X-ray beam through an individual who lies on the bed. A detector then measures the beam's attenuation and quantifies two components, these being boneless regions of fat and fat-free but mineral-free constituents, and bone mineral regions. The resultants are measures of **body fat** and of **bone mineral density**. The whole-body DEXA scan not only presents whole-body values but also regional composition as well. The six or seven regional components are head, torso, pelvis, and the four limbs. Since its development and use for determining % body fat, DEXA has become the new 'gold standard' measure because of its validity, reliability, and ease of use.

Bioelectrical impedance

Bioelectrical impedance is based on the electrical conductance characteristics of fat-free, hydrous, and fatty, anhydrous tissues. The impedance to the flow of an electrical current is a function of the resistance and reactance, and is related to the length and cross-sectional area of the fat-free, hydrous tissue. Electrodes are placed on an ankle and a wrist (on the same side of the body) and a small current passed between the electrodes. The resistance encountered is a function of the FFM, and hence determination of % body fat may be obtained. The system is easy to use, portable, and relatively cheap. Standardized conditions for its use must be adhered to, and these are to ensure that the body is hydrated when measurements are to be undertaken. Changes in hydration status of the body will affect the findings.

Near infrared reactance (NIR)

This technique uses a spectrophotometer and a fiber-optic probe which emits a beam of electromagnetic radiation in the near infrared region. The beam is directed at a site such as the biceps, and the returned, reflected energy is detected by a spectrophotometer. The beam penetrates muscle and subcutaneous fat before being reflected off bone, then being detected by the probe. It is not totally clear why or how body fat is determined using this method, although input of data concerning an individual's height, weight, sex, age, and level of activity are needed to allow computation of the results. NIR has not been used extensively and some results appear to demonstrate poor correlation with other measures of body fat.

J3 CHANGING BODY MASS AND BODY FAT

Key Notes

Reducing body fat

Obesity is a serious problem in many developed countries, and is linked with health-related diseases. Consequently, there is a need for obese individuals to reduce their body fat stores and so reduce the risk of diseases such as coronary heart disease and diabetes. Many athletes at various times of their training cycle increase body fat stores and subsequently have to reduce them before important matches.

Increasing lean body mass

The requirement to increase lean body mass is important for athletes who are underweight and those who engage in sports where being heavy is necessary for success. These sports are normally contact sports where a high power to weight ratio is important.

Related topics

Training for aerobic power (H2)
Physiological benefits of exercise (K2)

Guidelines for exercise prescription (L2)

Reducing body fat

The desire to lose body fat is common among competitive and recreational athletes and sedentary individuals. In some instances this need is necessary since the athlete may be above the required % body fat for their sport, and in the case of sedentary persons they may be overweight or even obese. Certainly from a health perspective being obese, and to a lesser extent being overweight, is a problem. There are sound epidemiological studies which have shown positive relationships between obesity and increases in incidence of coronary heart disease, strokes, hypertension, type II diabetes, and even some cancers. Equally, there are many studies establishing the requirement for athletes to have low % body fat for their chosen sports. How can body fat be reduced?

Weight loss, and in particular body fat loss, occurs when energy intake is less than energy expenditure. Exercise alone can be used to increase the energy expenditure and so result in weight loss. Likewise, reduced energy intake also leads to weight loss, as would a combination of diet and exercise.

Body fat loss through an exercise program is reliant on the fuels metabolized. Oxidation of fats occurs when the exercise intensity is moderate to low and when prolonged. *Fig. 1* shows how as exercise intensity increases there is a tendency to oxidize carbohydrates. It appears that the best exercise intensity for fat burning is around 65% $\dot{V}O_{2max}$. However, it must be remembered that greatest weight loss occurs when the energy expended throughout a day exceeds the intake from foods. Therefore, relatively intense exercise bouts will result in greater energy expenditure than lighter bouts. A sound compromise is to undertake prolonged moderate exercise, i.e. 30–60 minutes of aerobic exercise at a heart rate of around 75% of maximum.

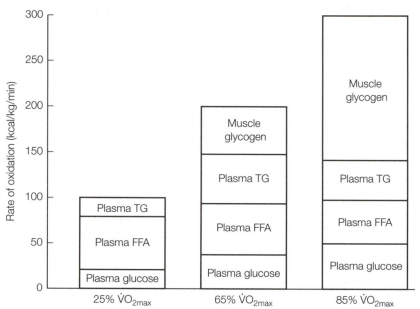

Fig. 1.   Energy sources in relation to exercise intensity. (TG = triglyceride; FFA = free fatty acid.)

Analysis of many studies, which have used exercise alone to reduce body weight, shows that the average loss is about 0.09 kg week^{-1}. This appears to be a small change for increased energy expenditure due to aerobic exercise. A problem is that some individuals increase their energy intake to nearly match the extra energy being expended, thus ensuring that the energy balance is only marginally negative. Therefore for exercise alone to be effective in reducing body weight, energy intake must be maintained at a level lower than the total daily energy expended. Compensatory eating is a problem. Another factor to consider is that exercise leads to an increase in FFM and a loss in body fat. Since muscle is denser than fat, a loss of body fat and a gain in muscle mass may even lead to a weight gain. Weight changes do not necessarily reflect changes in body fat.

Weight and body fat loss can also be achieved through dieting, i.e. reducing energy intake. Clearly the amount of food consumed when on a diet must be below the energy expended or else no weight will be lost. In addition, the composition of the food eaten may influence energy expenditure due to TEF. A high protein diet increases TEF and so may result in greater weight loss than a high fat or high carbohydrate diet of the same energy content. Health concerns with high protein, low carbohydrate diets such as the Atkins diet need to be addressed. In addition, there are nutritional benefits in consuming carbohydrates such as vegetables and fruit. The major consideration in dieting is that energy intake must be below energy expenditure, and that a variety of foods are eaten for health.

Research findings have clearly demonstrated the benefits of diet alone or exercise alone in reducing body weight. The process is normally slow with weight changes of around 0.09 kg week^{-1} for exercise alone and 0.5 kg week^{-1} for diet alone. It would appear that probably the most effective way to reduce body weight is by combining exercise and diet. A combination of diet and appropriate exercise should ensure that the energy balance is firmly negative, i.e. lower energy intake and greater energy expenditure. Indeed reports from studies in which both diet and exercise are used have resulted in mean weight loss between 1–1.5 kg week^{-1}. From a health and even a long-term perspective, diet and exercise are likely to be advantageous.

Increasing lean body mass

Weight gain is an issue for some athletes, particularly increase in lean body mass. The key to gaining weight is to consume more energy than is expended. Consuming a high-energy diet with a high carbohydrate, low fat, and adequate protein intake should lead to an increase in body weight. If the type of exercise is resistance and power orientated rather than aerobic, then the increase in body weight is likely to be directly attributed to muscle mass.

Eating more food to obtain the extra energy intake is best achieved by portioning out the meals rather than to eat more in two or three main meals. A 5 or 6 meal/snack regimen each day is preferable to a much larger three meal a day strategy. The emphasis on the type of foods is such that carbohydrates must contribute to a minimum of 60% energy intake with protein contributing to 15% of energy intake and fats approximately 25%. A careful and realistic body weight increase must be planned. Realistic weight gains of around 0.2–0.8 kg week^{-1} can be expected. If the amount of energy intake is necessarily high, liquid protein and carbohydrate supplements may be required in addition to meals.

K1 EXERCISE, FITNESS AND HEALTH

Key Notes

Physical activity vs physical fitness	Physical activity is defined as bodily movement or skeletal muscle contraction that results in energy expenditure. Physical fitness relates to a set of attributes that provide the individual with the ability to perform physical activity. It is imperative that these terms are used correctly in the field of exercise science.
Habitual physical activity	The amount of activity undertaken on a regular basis is termed habitual physical activity. A greater habitual activity level is associated with a decreased morbidity and mortality from chronic disease.
Current physical activity recommendations	Current recommendations for the optimal amount of physical activity for health gain recognize the importance of accumulation of activity, and of the intensity of activity undertaken.
Related topics	Exercise physiology (A1) Screening and exercise testing (L1) Energy balance (J1)

Physical activity vs physical fitness

Epidemiological evidence provides more and more data to suggest that inactivity, or a sedentary lifestyle is associated with an increase in **morbidity** and **mortality** from many chronic diseases. Diseases associated with lack of movement are termed **'hypokinetic.'** These diseases might include cancer, osteoporosis, obesity, coronary artery disease, etc. In order to understand how physical activity may be associated with health we need to be familiar with a number of definitions (see *Table 1*). These definitions are important because they

Table 1. Glossary of terms related to health related fitness.

Terminology	Definition
Physical activity	Any contraction of skeletal muscle that results in energy expenditure. Contraction does not always lead to movement.
Exercise	A subset of physical activity, which is volitional, planned, structured, repetitive and aimed at improving or maintaining fitness or health.
Sport	Physical activity involving structured and competitive situations, governed by rules.
Physical fitness	Relates to those components of fitness, i.e. strength, flexibility, cardiorespiratory efficiency, that are necessary to perform physical activity.
Physical health	A human condition with physical, social and psychological components. Positive health is associated with absence from disease and the capacity to enjoy life. Negative health is associated with morbidity and premature mortality.

distinguish between sport, exercise, physical activity and physical health. The exercise practitioner may succeed in providing an exercise program which enhances physical health, but does not improve physical fitness. Understanding this distinction is imperative when considering the effect of physical activity upon hypokinetic disease.

Habitual physical activity

The amount of physical activity undertaken on a regular basis by an individual is termed their **habitual activity level**. This level may include physical activity undertaken throughout the working day, but most often is measured as **leisure time physical activity** levels. **Epidemiological studies** attempt to associate amounts of physical activity with the incidence of disease, whilst **prospective studies** attempt to examine how increasing habitual physical activity influences the risk of, or the severity of, chronic disease. Given that physical activity increases energy expenditure, and that the more active a person is the more energy is expended, habitual physical activity levels are often recorded as the amount of energy expended. Energy expended is normally recorded in kcal or Joules and can be estimated through measures of oxygen consumption, or esti-mations of oxygen consumption from kilocaloric expenditure tables (see Section A). Habitual physical activity levels can thus be estimated by a number of means related to estimating energy expenditure (see also Section J1). These include:

- heart rate monitoring
- activity recall via interview or questionnaire
- motion sensor monitoring
- doubly labeled water assessment.

The simplest way to categorize habitual physical activity levels is to place individuals into one of three categories:

- Sedentary – Persons who are not accumulating 30 minutes or more of moderate-intensity physical activity on most days of the week.
- Moderately active – Persons who are accumulating 30 minutes or more of moderate-intensity activity on most days of the week. Moderate intensity refers to activity well within the individual's capacity, which can be sustained for a long period of time (usually 3–6 METs, 40–60% $\dot{V}O_{2max}$).
- Vigorously active – Persons who are participating in vigorous physical activity, greater than 6 METs or 60% $\dot{V}O_{2max}$.

Doubly labeled water is the gold standard for estimating energy expenditure as a function of physical activity. The technique allows for the measurement of the total energy expended over a 4–20-day period of time. The participant takes an oral dose of water containing a known quantity of **stable isotopes** (^{2}H and ^{18}O). As energy is expended these isotopes are lost in water and carbon dioxide, and the difference between the rate of loss of each isotope reflects carbon dioxide production. This can then be used to estimate energy expenditure using dietary information in the estimation. The doubly labeled water technique is accurate but is very expensive and requires specialized expertise.

Heart rate monitoring is an ideal way of assessing physiological responses to physical activity, and new monitors can easily record heart rate for a reasonable period of time. The very close relationship between heart rate and energy expenditure makes heart rate monitoring useful for assessing both energy expenditure and physical activity *per se*. Physical activity monitoring might involve assessing the amount of time spent in moderate-, or high-intensity activity. Monitoring of energy expenditure from heart rate measures requires the production of an individual heart rate–oxygen consumption calibration curve. This is because using a generalized relationship between heart rate and oxygen consumption to estimate energy expenditure would not take into account individual differences, and the effects of level of fitness. Consequently, energy expenditure can be estimated from recordings of heart rate using the **FLEX heart rate method**. The individual being assessed attends the laboratory in order to measure heart rate and oxygen consumption at rest whilst lying, sitting and standing, and then during exercise at a series of rising intensities. A heart rate–oxygen consumption curve is then produced, and the FLEX heart rate is quantified as the mean of the highest resting heart rate and the lowest exercise heart rate. When estimating energy expenditure over a period of time, resting metabolic rate is used if the recorded heart rate during activity falls below FLEX heart rate, and the calibration curve is used if the recorded heart rate during activity falls above FLEX heart rate. Although the calibration curve is only really relevant to the specific activity undertaken, the FLEX heart rate method has been seen to fall within 10–18% of the values recorded using the doubly labeled water technique.

Physical activity questionnaires or interviews are wholly dependent upon individual recall, however they are cost effective and ideal for large population studies. There are a number of questionnaires available to use in the field, and it is imperative to use a population-specific design. The **Baecke questionnaire** assesses activity undertaken at work, in sport and in leisure time, and then provides a total activity index for young adults. The questionnaire is reliable and has been validated against a number of other self-report questionnaires, and motion sensors. Other questionnaires include the **Tecumseh questionnaire**, the **Minnesota Leisure Time Physical Activity Questionnaire** (MLTPA), the **Physical Activity Scale for the Elderly** (PASE), and the **Physical Activity Score** for children to name a few. These questionnaires allow for the calculation of kilocaloric expenditure over differing periods of time, usually 7–14 days, based on laboratory measurements of the kilocaloric expenditure of numerous activities. These questionnaires are reliable and many have been validated against a number of other self-report questionnaires, and motion sensors, whilst the Baecke, Tecumseh, MLTPA and PASE questionnaires have been validated against the doubly labeled water technique.

Motion sensor monitoring includes the use of pedometers and accelerometers. **Pedometers** allow for the measurement of the number of steps taken over the measurement period. They achieve this by monitoring vertical displacement of a specific object, for example the belt upon which it is attached. It seems from a number of studies that the pedometer will give different recordings when worn on the belt, than if worn for example on the ankle, and again if an individual walks slowly, at normal pace or very quickly. If walking ability is the measure required the pedometer is an accurate and cost-effective tool, however as it is not sensitive enough to detect stride length or total body displacement, it is not a valid tool for assessing energy expenditure over time. **Accelerometers** electronically detect total body displacement either in a single plane (**uniaxial**) or in three planes (**triaxial**) of motion.

Uniaxial accelerometers such as the Caltrac and CSA, are designed to monitor accelerations and decelerations of the body mass, and to estimate energy expenditure from these displacements. The reliability of unaxial accelerometers is good, however validity is questioned for several reasons. The energy expended during gait is not completely reflected in the acceleration and deceleration of body mass, and indeed the energy expended during limb rotation, isometric and eccentric exercise or load carriage would not be identified by the accelerometer. When uniaxial accelerometers are compared to laboratory calorimetry measures of energy expenditure they appear to overestimate energy expenditure, whilst underestimating energy expenditure when compared to the doubly labeled water technique.

Triaxial accelerometers such as the Tritrac R3D and the Tracmor, measure acceleration in the vertical, horizontal and mediolateral planes. Although the triaxial accelerometers seem to have a better comparison to laboratory measures of energy expenditure than the uniaxial, they still appear to underestimate energy expenditure when compared to the doubly labeled water technique. It thus appears that accelerometers are useful in comparing physical activity levels between groups, or before and after an exercise intervention, but may not be accurate in establishing energy expenditure.

Current physical activity recommendations

Based upon the understanding that an active lifestyle decreases the mortality and morbidity associated with disease, and on the understanding of how exercise physiologically influences specific diseases (see Section K3), there are a number of organizations recommending a more active lifestyle. More specifically the **UK Health Development Agency**, the **ACSM** and **US Centers for Disease Control** recommend that: 'Adults should accumulate 30 minutes or more of moderate-intensity physical activity on most, preferably all, days of the week.'

Moderate-intensity activity in this statement referred to activities that use approximately 150 kilocalories (630 kJ) per day or are equivilent to 40–60% of $\dot{V}O_{2max}$. The statement also highlights that many health benefits may be accrued by **accumulating** short bouts of activity throughout the daytime. This recommendation should be seen as the minimal recommendation for health benefit as not all diseases respond to moderate-intensity activity. Indeed, the Surgeon General of the United States later updated the recommendation to state that: 'Additional health benefits can be gained through greater amounts of physical activity. People who can maintain a regular regimen of activity that is of longer duration or of more vigorous intensity are likely to derive greater benefit' (US Surgeon General Report, 1996).

K2 PHYSIOLOGICAL BENEFITS OF EXERCISE

Key Notes

Overview	Rather than reviewing all of the physiological effects of exercise, this section points the student to the potential benefits of exercise in the form of a diagram.
Related topics	Sections A–L

Overview

The physiological responses to chronic physical activity have been reviewed in the previous chapters. *Fig. 1* attempts to highlight these responses and to link them to the benefits identified for increasing health, and decreasing morbidity and mortality. This figure can then be compared to each of the chapters in this book.

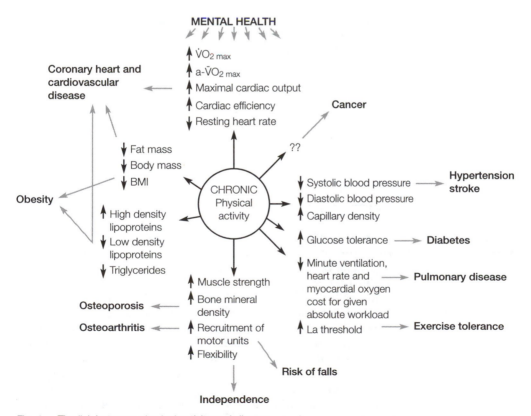

Fig. 1. The link between physical activity and disease.

K3 EXERCISE AND DISEASE

Key Notes

Cardiovascular and coronary heart disease
Physical activity has been associated with a protective effect for cardiovascular and coronary heart disease. The effectiveness of vigorous activity is however unclear.

Stroke
Strokes are associated with ischemia in the cerebrovasculature. Physical activity has not been directly associated with protection against a stroke, however the effect of exercise upon risk factors for cardiovascular disease are thought to indirectly reduce the risk of suffering from a stroke.

Cancer
The evidence for any protective effect of physical activity for cancer is contradictory. However, the risk of colon cancer and breast cancer may be lessened by participation in physical activity.

Diabetes mellitus
The major types of diabetes are insulin-dependent diabetes mellitus (IDDM) and non-insulin-dependent diabetes mellitus (NIDDM). Both types of diabetes respond well to exercise.

Osteoporosis
Osteoporosis is characterized by a low bone mineral density and altered structural integrity. Physical activity is associated with an increased bone mineral density both in prevention and treatment of osteoporosis.

Osteoarthritis
Osteoarthritis is characterized by a degeneration of cartilage around joints. Physical activity has been thought to both protect against and to cause osteoarthritis although convincing evidence for either argument is lacking. Physical activity should be utilized with sufferers to aid in reducing pain and inflammation around joints.

Related topics
Physiological benefits of exercise (K2)

Special populations (L3)

Cardiovascular and coronary heart disease

The early classic epidemiological studies associating physical activity and cardiovasclar disease were conducted in the 1950s and 1960s in England and the USA on London bus conductors and San Francisco longshoremen. Jerry Morris found that the greater physical fitness seen in bus conductors than in bus drivers and civil servants was associated with a lower incidence of, and lower early mortality from, coronary heart disease (CHD). Paffenbarger and colleagues also found a lower incidence and mortality from coronary heart disease in the most active longshoremen. Paffenbarger went on to study the influence of habitual and leisure time physical activity upon coronary heart disease in 16 936 Harvard University alumni between 1962 and 1972. Once again, at all ages mortality was lower in the active alumni. These findings have been reiterated, especially with respect to walking by studies such as the **Nurses Health Study** and the **Framingham Heart Study**.

The **US Railroad Study** evaluated death from CHD and all-cause mortality in railroad workers via a 17–20-year follow-up. The relative risk of CHD death was 1.3 times greater in men who had expended 40 kilocalories per week compared to men who had expended 3632 kilocalories per week. These risk factors had been adjusted for differences in age, smoking, blood pressure and serum cholesterol, and indicated the effectiveness of both moderate and vigorous physical activity in protecting against disease states. However, the **British Regional Heart Study** investigated the relationship between physical activity levels and the risk of CHD and stroke in middle-aged British men in 24 towns. These towns represented the spread of socio-economic groups. Energy expenditure was divided into inactive, occasional, light, moderate, moderately vigorous, and vigorous. During the follow-up period 488 men suffered at least one coronary event, however the risk of CHD decreased with increased physical activity; the groups reporting moderate or moderately vigorous activity experienced less than half the rate for inactive men. Vigorously active men, however, experienced higher rates of CHD along with those describing themselves as occasionally or lightly active. The authors concluded that **moderate physical activity** is associated with lower rates of CHD and stroke in men both with and without pre-existing CHD, and that this relationship remained over an 8-year follow-up period. This upturn in risk associated with **vigorous activity** has also been reported by the **Multiple Risk Factor Intervention Trial (MRFIT)**, the **Harvard Alumni Study** and the **Copenhagen Male Study**.

Stroke

Strokes are **cerebrovascular events** usually associated with ischemia in blood vessels supplying the brain. This ischemia leads to nerve death and thus damage to speech, sight and behavior and often paralysis. Risk factors associated with stroke are high blood pressure, heart disease, cigarette smoking, high red blood cell count and **transient ischemic attacks** (TIAs). A TIA is defined as sudden focal loss of neurological function, caused by inadequate perfusion, with complete recovery usually within 24 hours.

The majority of studies have not indicated a relationship between habitual physical activity level and stroke. It may be that as the risk factors for stroke are similar to those for coronary artery disease, modifications in blood pressure especially might decrease the risk of hemorrhagic stroke. A recent study by Kurl *et al.* (2003) reported that men with low cardiorespiratory fitness at the start of an 11-year study were three times more likely to suffer a stroke than men with a $\dot{V}O_{2max}$ of greater than 35.3 ml kg min^{-1}. This association remained after adjusting for a number of other risk factors.

Cancer

Paffenbarger's Harvard Alumni study also indicated an increased risk of cancer with a sedentary lifestyle whilst other studies have found no convincing relationship between physical activity and cancer. The most convincing evidence linking cancer risk to habitual physical activity seems to be for **colon cancer**. As activity levels increase the risk of colon cancer decreases, with sedentary individuals having approximately double the risk compared to active. It may be that physical activity decreases the risk of colon cancer by increasing intestinal motility via the production of prostaglandins, and by decreasing gastrointestinal transit time. Some relationships have been reported between physical activity and breast cancer, testicular cancer, rectal cancer and prostate cancer but the evidence is often contradictory in nature. It does, however, seem that women active in moderate to vigorous exercise may gain a decreased risk of breast cancer, although this relationship is reversed if obesity is present. Further studies in these areas are called for, especially where diet and activity can be assessed together.

Diabetes mellitus   The major types of diabetes are **insulin-dependent diabetes mellitus** (IDDM) and **non-insulin-dependent diabetes mellitus** (NIDDM). IDDM is caused by destruction of the β cells that produce insulin in the pancreas. As insulin deficiency influences the rate of glucose uptake in the blood the disease is controlled through regular insulin injections. NIDDM, the more common form of diabetes, is caused by a reduced sensitivity of the target cells for insulin, again influencing glucose regulation. The majority of NIDDM sufferers are obese, and thus this disease is often managed via diet, weight loss and exercise.

Sufferers of IDDM have been seen to benefit from exercise in terms of improving coronary artery disease risk and insulin sensitivity of cells, however no benefits have beeen seen in glucose regulation following an exercise intervention. Exercise for NIDDM sufferers has been seen to decrease blood cholesterol and triacylglyerol concentrations, decrease weight and improve glucose regulation. Blood glucose levels should be monitored prior to exercise sessions and NIDDM sufferers are recommended to exercise for 40–60 minutes in duration, 3–5 times per week, whilst IDDM sufferers should reach 20–60 minutes daily.

Osteoporosis   **Osteoporosis** is characterized by a decreased **bone mass** and structural deterioration of bone tissue leading to bone fragility. The disease increases the risk of fractures, especially on falling, and is associated with the aging process. With this in mind maintenance of bone mass, especially in **post-menopausal women** who lose the protective effects of the hormone estrogen, is imperative. A physically active lifestyle is associated with a decreased risk of osteoporosis because of the mechanical effect of exercise upon bone tissue. **Forces** placed upon bone in an overloading fashion cause bone remodeling in order to enhance bone mineral density, and thus bone strength. Inactivity leads to resorption of bone and a decreased density and strength. What should be remembered when using exercise to prevent osteoporosis, or to enhance existing low bone density, is that bone will only respond at the specific site at which it is loaded. Consequently, bone responds best to loads of high magnitude and rate applied in multiple directions.

Osteoarthritis   **Osteoarthritis** is the most common arthritis and is characterized by degeneration of **cartilage** and growth of new bone around the joint. Incidence of the disease increases with age and is a common cause of inactivity in elderly persons. Historically, whether physical activity is associated with protection against osteoarthritis or is a cause of arthritis has been contentious. Indeed, some cross-sectional studies have suggested a greater incidence of osteoarthritis in cohorts of people who have been physically active, and especially in **sports** participants. To date there is still no real understanding of how activity influences the risk of the disease, especially as injury *per se* may be the causative factor in sports participants. For those people with osteoarthritis and **rheumatoid arthritis** regular aerobic exercise and resistance training has been associated with a decrease in pain, reduction in joint swelling and an increased cartilage production. High-intensity exercise and repetitive high impact are not recommended for arthritis sufferers.

L1 SCREENING AND EXERCISE TESTING

Key Notes

Preparticipation screening

All individuals who wish to take part in an exercise program, or in an exercise test, should be screened. This screening involves an evaluation of current and past medical status, and thus allows the assessment of health status.

Clinical evaluation

As part of the medical screening individuals should be examined for a number of variables that indicate health status. These include factors such as body mass, blood pressure, blood cholesterol and glucose concentrations, pulmonary function and signs and symptoms of disease. These variables can also be used as baseline measures for the start of an exercise program.

Risk stratification

A medical history, physical examination and clinical evaluation provide the detail required in order to classify an individual into a risk stratum. The risk stratification provides an index of the risk of having an untoward event during physical activity, and as such aids the practitioner in developing an appropriate exercise prescription.

Physical fitness testing

In developing an appropriate exercise prescription knowledge of the current level of fitness of the client is imperative. Fitness testing usually concentrates upon parameters of health-related fitness. These include body composition, cardiorespiratory fitness, muscular fitness and flexibility.

Related topics

Estimation and measurement of energy expenditure (A4)
Exercise, fitness and health (K1)

Guidelines for exercise prescription (L2)

Preparticipation screening

Prior to any individual participating in any exercise program, testing or prescription, it is imperative that they are **health screened**. Irrespective of the legal implications for the exercise professional and/or the exercise facility, the safety of the client or subject must be paramount. The screening process thus acts as a tool to identify the **current health status** of the participant, and consequently the potential risks inherent in any further exercise testing or prescription. The screening process should include the establishment of a full medical history, inclusive of both past and present information. The **components of a medical history** should relate to: (a) any medical diagnosis in existence, e.g. cardiovascular disease, osteoporosis etc; (b) any other previous physical examination results, e.g. high blood pressure, abnormal blood sugar etc.; (c) any history of symptoms of disease, e.g. shortness of breath, dizziness, etc.; (d) any orthopedic problems; (e) any family history of cardiac, pulmonary or metabolic disease; (f) any recent hospitalization or illness; (g) the use of any medications; (h) the use of alcohol, tobacco, caffeine or drugs; (i) the physical demands of current occupation; and (j) exercise history and current habitual physical activity. The minimal standard for entry into moderate intensity programs has been recommended to be the **Physical Activity Readiness Questionnaire (PAR-Q)**. This questionnaire provides a broad-brush approach to the information needed from the medical history in a manner that should allow for the identification of persons needing medical approval to exercise. Details on the PAR-Q can be found in ACSM (2000) and Heyward (1998).

Clinical evaluation

A number of basic tests, conducted at rest, are useful in undertaking a clinical evaluation of the participant prior to participation in exercise tests or prescription. The results of these tests are added to the knowledge gained in the medical history in order to build a comprehensive picture of the current health status of the participant. They are also obviously useful as baseline measures to compare to following an exercise program. These tests should include resting **blood pressure** (taken on at least two separate occasions), **blood cholesterol** and **lipoprotein concentrations, fasting blood glucose, body mass, waist and hip circumference**, and potentially **serum triglycerides, pulmonary function, body composition analysis** and a **physical examination**.

Blood pressure should be recorded as both systolic blood pressure and diastolic blood pressure (see Section E). These measures can then be used to classify the client as having normal or abnormal levels of blood pressure. The cut off points for each category can be seen in *Table 1*. **Blood cholesterol, lipoprotein** and **triglyceride concentrations** likewise can be classified into low, normal and high categories (*Table 2*).

Table 1.   Blood pressure classification for persons > 18 years

Classification	Systolic BP (mmHg)	Diastolic BP (mmHg)
Normal	<130	<85
High normal	130–139	85–89
Mild hypertension	140–159	90–99
Moderate hypertension	160–179	100–109
Severe hypertension	180–209	110–119
Very severe hypertension	≥210	≥120

Data from Sixth Report of the Joint Committee on Prevention, detection, evaluation, and treatment of high blood pressure, National Institutes of Health, National Heart, Lung and Blood Institute, NIH Publication No. 98-4080, November, 1997.

Body composition analysis provides an evaluation of percentage body fat and fat-free mass (see Section J for further details). These values are best used as an index of **obesity** and as an important tracking variable if the client is interested in a weight-loss program. This is especially important if body mass is stable, but body composition has changed substantially. The **waist:hip ratio** is also an important epidemiological variable used as an index of obesity. *Fig. 1* indicates how body fat is generally stored in men and women, how this is evaluated and what the ratio might mean. Additionally, the ACSM (2000) stipulate that a **waist circumference** of greater than 100 cm, or a **body mass index** of equal to, or greater than, 30 kg m^2 acts as an independent risk factor for coronary artery disease.

Table 2. Classification of total cholesterol, triglycerides and lipoproteins. Data in mg dL^{-1}, brackets are mmol L^{-1}

Classification	Total cholesterol	LDL-C	Triglcerides
Desirable	<200 (<5.2)	<130 (<3.4)	<200 (<2.26)
Borderline high	200–239 (5.2–6.2)	130–159 (3.4–4.1)	200–399 (2.26–4.51)
High	≥240 (≥6.2)	≥160 (≥4.1)	400–999 (4.52–11.28)
Very high	–	–	≥1000 (≥11.29)

Classification	HDL-C
Low	<35 (<0.9)
Normal	35–60 (0.9–1.5)
High	>60 (>1.5)

Data from National Cholesterol Education Program Committee (1993) Summary of the second report of the Cholesterol Education Program (NCEP) Expert Panel on Detection, Evaluation, and Treatment of High Blood Cholesterol in Adults. *J. Am. Med. Ass.* 269: 3017.

Body mass index (BMI) is calculated as:

$$BMI = \frac{body\ mass\ (kg)}{height^2\ (m)}$$

So for example an individual with a body mass of 65 kg and a height of 164 cm would have a BMI of:

$$BMI = \frac{65}{1.64^2} = 24.17\ kg\ m^{-1}$$

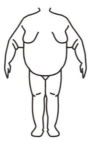

Waist circumference measured at the level of the umbilicus

Hip circumference measured at the level of the greatest protrusion of the buttocks

Both circumferences measured from the side of the client with the client standing with feet together

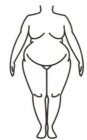

Apple shape
Android obesity
Greater health risks
associated with
W : H>0.94

Pear shape
Gynoid obesity
Greater health risks
associated with
W : H>0.82

Fig. 1. The waist : hip ratio as an index of general health and fat patterning.

Care should be taken when utilizing the BMI as a measure of fatness because many people may be heavy for their height due to a dense skeleton or greater muscle mass. In these people, rugby players or rowers for example, the BMI will indicate that they are over fat. Consequently, the BMI should be used with some common sense based upon its utility with individuals who are obviously over fat.

Risk stratification The combined knowledge of the medical history and the basic clinical evaluation aids the professional to place the participant into a **risk stratum** based on the risk for the development of **coronary artery disease**, and/or **signs and**

symptoms of known **cardiovascular, pulmonary** or **metabolic disease** (see *Tables 3* and *4*). Knowing which stratum the participant falls into allows the professional to make relevant decisions regarding: (a) the need for exercise testing prior to exercise program participation; (b) the need for medical supervision of exercise testing; and (c) what type and intensity of exercise is appropriate for

Table 3. Coronary artery disease risk factor (thresholds for use with ACSM risk stratification*) Reprinted with permission from ACSM's Guidelines for exercise testing and prescription, sixth edition, 2000.

Risk factors	Defining criteria
Positive	
Family history	Myocardial infarction, coronary revascularization, or sudden death before 55 years of age in father or other male first-degree relative (i.e., brother or son), or before 65 years of age in mother or other female first-degree relative (i.e., sister or daughter)
Cigarette smoking	Current cigarette smoker or those who quit within the previous 6 months
Hypertension	Systolic blood pressure of ≥140 mmHg or diastolic ≥90 mmHg, confirmed by measurements on at least 2 separate occasions, or on antihypertensive medication
Hypercholesterolemia	Total serum cholesterol of >200 mg/dL (5.2 mmol/L) or high-density lipoprotein cholesterol of <35 mg/dL (0.9 mmol/L), or on lipid-lowering medication. If low-density lipoprotein cholesterol is available, use >130 mg/dL (3.4 mmol/L) rather than total cholesterol of >200 mg/dL
Impaired fasting glucose	Fasting blood glucose of ≥110 mg/dL (6.1 mmol/L) confirmed by measurements on at least 2 separate occasions
Obesity†	Body Mass Index of ≥30 kg/m², or waist girth of >100 cm
Sedentary lifestyle	Persons not participating in a regular exercise program or meeting the minimal physical activity recommendations‡ from the US Surgeon General's report
Negative	
High serum HDL cholesterol§	>60 mg/dL (1.6 mmol/L)

*Adapted from Expert Panel on Detection, Evaluation, and Treatment of High Blood Cholesterol in Adults. Summary of the second report of the National Cholesterol Education Program (NCEP) expert panel on detection, evaluation, and treatment of high blood cholesterol in adults (Adult Treatment Panel II). JAMA 1993; 269: 3015–3023.

†Professional opinions vary regarding the most appropriate markers and thresholds for obesity; therefore, exercise professionals should use clinical judgment when evaluating this risk factor.

‡Accumulating 30 minutes or more of moderate physical activity on most days of the week.

§It is common to sum risk factors in making clinical judgments. If high-density lipoprotein (HDL) cholesterol is high, subtract one risk factor from the sum of positive risk factors because high HDL decreases CAD risk.

Table 4. Signs and symptoms of cardiovascular, pulmonary and metabolic disease

Cardiovascular	Pulmonary	Metabolic
Heart murmurs	Asthma	Diabetes
Palpitations/tachycardia	Shortness of breath with	Obesity
Ischemia/angina	mild exertion	Thyroid disorders
Previous MI	Bronchitis	Renal disease
Previous stroke	Emphysema	Liver disease
Dizziness/fainting	Nocturnal dyspnea/difficulty	Glucose intolerance
Ankle edema	breathing	McArdle's syndrome
Chest/neck/arm/jaw pain	Coughing up blood	
Shortness of breath		
Hypertension		
Hypercholesterolemia		
Claudication/calf pains		

the participant. The most popular risk factor profile used for initial profiling is the **American College of Sports Medicine Risk Stratification** (ACSM, 2000). This stratification places the participant into a stratum of either low, moderate or high risk of the occurrence of an untoward event happening during exercise. The ACSM states that:

- Low-risk individuals do not show signs and symptoms of cardiovascular or pulmonary disease and have no more than one risk factor for coronary artery disease.
- Moderate-risk individuals display two or more risk factors for coronary artery disease OR are men 45 years or above, OR women 55 years or above.
- High-risk individuals show signs or symptoms of cardiovascular or pulmonary disease or have a diagnosed cardiovascular, pulmonary or metabolic disease.

The integration of knowledge from screening and risk stratification can be utilized to aid in the construction of the exercise prescription. **High-risk** individuals should be examined by a medic, and should undergo exercise testing with medical supervision before participation in any exercise program (ACSM, 2000). This advice is also given for **moderate-risk** individuals who wish to participate in vigorous (very hard) intensity activity. Consequently the results of screening might be:

- accept client
- refer client for further medical screening prior to any exercise prescription
- refer client to a facility with more appropriate levels of support.

Physical fitness testing

Knowledge of parameters of health-related physical fitness allows the practitioner to gain an understanding of the client's baseline fitness status, to compare this to published norms and to construct achievable goals for the exercise prescription. **Cardiorespiratory fitness** is usually assessed using the concept of $\dot{V}O_{2max}$ (see Section A), and this can be estimated through either a **maximal** or **submaximal** exercise test. Although maximal exercise testing provides the greatest sensitivity for the diagnosis of coronary artery disease, it is not always practical or achievable in the exercise facility. In these cases submaximal exercise tests provide a practical, cost-effective and reasonably accurate reflection of a client's cardiorespiratory fitness. An example of the types of tests that can be used to estimate $\dot{V}O_{2max}$ can be seen in *Table 5*.

The commonly used treadmill and cycle ergometer tests employ a series of rising exercise intensities (increments). These tests are termed **incremental** or **graded exercise tests** whereby treadmill speed, treadmill gradient or the load applied to the ergometer is increased every 2–3 minutes dependent upon the protocol chosen. Choice of the protocol is very much dependent upon the baseline fitness/health of the client, and many **stress tests** used in hospital laboratories involve walking protocols with increasing treadmill gradient. In **maximal exercise tests** it is usual to measure heart rate, oxygen consumption, blood pressure and the rating of perceived exertion (RPE) at the end of every increment until volitional exhaustion. In **submaximal exercise tests** oxygen consumption is predicted from an assumed linear relationship between heart rate and oxygen consumption. The client undertakes a limited number of increments and heart rate is recorded at the end of each increment. These heart rates are plotted on a graph as in *Fig. 2* and the heart rate response is extrapolated

Table 5.   Tests used for estimating $\dot{V}O_{2max}$

Mode of testing	Example of tests used
Field tests	Cooper 12 minute distance test
	1.5 mile test for time
	Rockport 1 mile fitness walking test
	Shuttle test
Treadmill tests	British Association of Sport and Exercise Sciences Maximal test
	Bruce, Balke, Naughton, Stanford protocols
Cycle ergometer tests	British Association of Sport and Exercise Sciences Maximal test
	Astrand-Rhyming protocol
	YMCA protocol
	PWC_{170} protocol
Step tests	3 min YMCA test
	Canadian home fitness test
	Astrand-Rhyming test
	Maritz protocol

upwards to meet the age-predicted maximal heart rate (220 – age). Oxygen consumption is then predicted by dropping a line downwards to predict the work rate that would have been reached at this heart rate. An example is provided for a 40-year-old 70-kg person in *Fig. 2*. In the figure the predicted maximal work rate was 875 kg m^{-1} min^{-1}. The ACSM (2000) prediction equations for oxygen consumption are then utilized to predict oxygen consumption at this work rate.

The **ACSM (2000) prediction equations** are different for walking, treadmill and outdoor running and cycle ergometry but all predict oxygen consumption (in ml kg^{-1} min^{-1}) for any speed, work rate or gradient. For the example above the leg or cycle ergometry prediction equation states:

$$\dot{V}O_2 = \left(\frac{10.8 \times W}{M}\right) + 3.5 \quad \text{where W is power in Watts and M is body mass in kg}$$

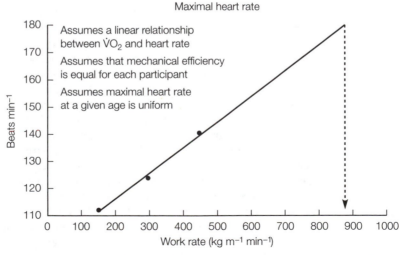

Fig. 2.   A plot of heart rates recorded at the end of three increments of a submaximal exercise test to predict $\dot{V}O_{2max}$. Section L1 explains how to predict $\dot{V}O_{2max}$ from this plot.

So we know from the submaximal test:

(i) Body mass = 70 kg

(ii) Predicted maximal work rate was 875 kg m^{-1} min^{-1}

What we don't know is the power the client was working at during the test. However, as reviewed in Section A we know that 1 Watt is equal to 6.12 kg m^{-1} min^{-1}, so 875 kg m^{-1} min^{-1} = 143 Watts (875/6.12).

So we can now complete the equation:

$$\dot{V}O_2 = \left(\frac{10.8 \times 143}{70}\right) + 3.5$$

So, if the client had been cycling at 875 kg m^{-1} min^{-1} as predicted from the submaximal cycle ergometer test, his $\dot{V}O_2$ would have been:

25.56 ml kg^{-1} min^{-1}

As the client had a body mass of 70 kg this would mean a predicted maximal oxygen uptake of:

25.56 ml kg^{-1} min^{-1} × 70 kg = 1789 ml OR 1.79 litres per minute

As 1 MET = 3.5 ml kg^{-1} min^{-1} this client has a **functional capacity** of:

25.56 ml kg^{-1} min^{-1} ÷ 3.5 ml kg^{-1} min^{-1} = 7 METs

The equations for walking, running and stepping can be found in ACSM (2000).

The assessment of **muscular fitness** integrates both **muscular strength** (the maximal force a muscle can generate at a specific velocity) and **muscular endurance** (the ability of a muscle to resist fatigue during repeated contractions). Muscle strength can be measured on many isokinetic and isometric dynamometers and tensiometers, however as the baseline measure is usually used to determine training thresholds a dynamic test is warranted. The most favored approach is to measure a **1 repetition maximum (1-RM)**, or the most weight a client can lift once only with good form. The 1-RM is usually found by asking the client to lift 60–80% of their perceived maximal lift, and then to add small weights until the true 1-RM is reached. Many practitioners use the 8 or 12-RM to evaluate muscular fitness, as this can be tracked more easily throughout a weight-lifting exercise program. Standardized muscular endurance tests tend to be the total number of **push-ups** possible, the **Canadian curl-up test**, or the number of **sit-ups** possible in 1 minute. Norm values and more details of the standardization procedures for these tests can be found in ACSM (2000) and Heyward (1998).

Flexibility, or the ability of a joint to move through its full range of motion, is another physical fitness test indicative of general fitness and functional ability. The assessment of flexibility should only be undertaken after a suitable warm up, and should identify those areas needing attention. Flexibility can be assessed with the use of **goniometers** that record the **range of motion (ROM)** a joint can achieve. The **sit and reach test** is the most commonly used test of flexibility, however it should be recognized that this test only evaluates low back and hip flexibility, and that even then it may reflect hip flexibility to a much greater extent than low back flexibility. Nevertheless, flexibility in joints such as these plays a role in reducing pain from some joint problems, and in increasing function, especially in activities of daily living (e.g. getting out of the bath). Details of flexibility norms and procedures can be found in ACSM (2000) and Heyward (1998).

Following the screening and physical fitness testing the practitioner has a detailed record of the client's present goals and condition. This record can be used as a baseline analysis of condition prior to an exercise program, and to aid in writing the exercise prescription using the guidelines reviewed in Section L2.

L2 GUIDELINES FOR EXERCISE PRESCRIPTION

Key Notes

Principles of training

The principles of training are based upon overload and specificity.

The training session

The training session should include a warm up, a stimulus session, and a cool down.

Exercise intensity

Exercise intensity is usually calculated as a percentage of an individual's maximal capacity. Thus an exercise program may be written to include exercise conducted at a percentage of maximal heart rate, maximal oxygen consumption, heart rate or oxygen consumption reserve or one or 10–12 repetition maximum.

Exercise duration

The duration of an exercise session or activity is inversely related to exercise intensity and controls the caloric expenditure of the session.

Exercise frequency

The number of times per week an exercise session is conducted influences the total caloric expenditure of the activity, the incidence of injury and adherence to an exercise program.

Rate of progression

The rate of progression depends upon the initial fitness level of the participant. There are three stages to progress through: (i) the initial stage; (ii) the improvement stage; and (iii) the maintenance stage. The participant's goals and physiological parameters of health-related fitness should be reassessed throughout each stage.

Related topics

Estimation and measurement
of energy expenditure
(A4)

Training for performance
(Section H)
Screening and exercise testing (L1)

Principles of training

The principles of training (see also Section H) for health-related fitness are the same as for physical fitness *per se*. Improvements in cardiorespiratory fitness, muscular fitness and flexibility are dependent upon the principles of **overload** and **specificity**. That is to say that training a specific muscle for increased strength will not have a large impact upon cardiorespiratory fitness – the training was specific to increasing strength! Furthermore any increase in muscular strength in this example is dependent upon the muscle being worked at a load that is greater than that to which it is normally accustomed. Working against a low resistance with high repetitions is likely to increase muscular endurance but not muscular strength. What is important to remember when using exercise to increase **health-related fitness** is that the overload might be exceedingly small, and may indeed take time to reach. Working with individuals of very low functional capacity, and/or with specific diseases dictates a slow and careful approach to overload. **Furthermore, it should be recognized that a very small increase in one of the components of health-related physical fitness might relate to a huge increase in physical health.**

The training session

In designing an **exercise prescription** the components of the exercise program as a whole, and the individual training session need to be addressed. The exercise program should include cardiorespiratory and muscular fitness elements, with the inclusion of recreational activities to boost enjoyment and adherence. The prescription should be designed to meet the goals of the participant with the main aim of increasing habitual physical activity. An individual training session should include a 10-minute **warm up** period, a stimulus session that might be endurance or resistance training, followed by a 5–10-minute **cool down**. The warm up period is important to increase metabolic rate and body temperature, to stretch postural muscles and to decrease the risk of injury. It should aim to progressively reach a level just below what is expected in the exercise session, so that a fast walking exercise prescription should warm the client up to a slow walk. A jogging session should warm the client up to reach a fast walk. The cool down is imperative to aid the clearance of waste metabolites built up during exercise, to allow for the dissipation of heat and to ensure appropriate circulatory adjustments that decrease the risk of cardiac arrhythmias that often precede sudden cardiac death.

Exercise intensity

Exercise intensity (see also Section A4) provides a measure of how hard the client is working relative to their maximal capacity for the given activity. As a consequence intensity is inversely related to exercise duration, i.e. as intensity increases duration normally decreases. Exercise intensity for **cardiorespiratory work** is usually expressed as a percentage of the individual's:

$\dot{V}O_2$ max
$\dot{V}O_2$ **reserve** ($\dot{V}O_2R$)
maximal heart rate (HR_{max}) or
heart rate reserve (HRR)

$\dot{V}O_2R$ and HRR refer to the difference between $\dot{V}O_{2max}$ or HR_{max} and resting values. The measurement or prediction of maximal $\dot{V}O_2$ or maximal heart rate was addressed in Section L1.

The ACSM have produced general guidelines indicating the **target intensity** for cardiorespiratory training (ACSM, 2000). These guidelines are provided in *Table 1*. These targets provide a broad band of intensities through which the practitioner needs to reflect on the initial condition of the client. So a deconditioned person may benefit from exercising at an intensity of 40–50% HRR, whereas a client already involved in physical activity may need to exercise at 60–80% HRR to gain any benefit. So for example, a 24-year-old individual (resting heart rate of 70 beats min^{-1} and $\dot{V}O_{2max}$ of 20 ml kg^{-1} min^{-1}) needing to exercise at an intensity of 70–85% HR_{max} would have a target heart rate of:

Maximal heart rate: $220 - 24 = 196$ beats min^{-1}
Lower target = 70% of 196 = 137 beats min^{-1}
Upper target = 85% of 196 = 167 beats min^{-1}

The same individual needing to exercise at an intensity of 60–80% HRR would need to exercise at a target heart rate of:

Table 1.   ACSM (2000) recommended exercise intensities for improving cardiorespiratory fitness

Method of measurement	Recommended exercise intensity
$\dot{V}O_{2max}$	55/65 – 90%
$\dot{V}O_2$ reserve	40/50 – 85%
HR_{max}	55/65 – 90%
HR reserve	40/50 – 85%
METS	Corresponds to $\dot{V}O_2$ reserve

Maximal heart rate: $220 - 24 = 196$ beats min^{-1}
Resting heart rate: 70 beats min^{-1}
Maximal heat rate – resting heart rate
$= 196 - 70 = 126$ beats min^{-1}

Lower target $= 60\%$ of $126 = 76$ beats min^{-1}
Add back on resting component $= 76 + 70 = 146$ beats min^{-1}
Upper target $= 80\%$ of $126 = 101$ beats min^{-1}
Add back on resting component $= 101 + 70 = 171$ beats min^{-1}

Target heart rate for exercise session $= 146 - 171$ beats min^{-1}

The same individual needing to exercise at an intensity of 60–80% $\dot{V}O_2R$ would need to exercise at a target $\dot{V}O_2$ of:

Maximal – resting $\dot{V}O_2 = 20 - 3.5 = 16.5$ ml kg^{-1} min^{-1}
Lower target $= 60\%$ of $16.5 = 9.9$ ml kg^{-1} min^{-1}
Add back resting component $= 9.9 + 3.5 = 13.4$ ml kg^{-1} min^{-1}
Upper target $= 80\%$ of $16.5 = 13.2$ ml kg^{-1} min^{-1}
Add back resting component $= 13.2 + 3.5 = 16.7$ ml kg^{-1} min^{-1}

Target $\dot{V}O_2$ for exercise session $= 13.4–16.7$ ml kg^{-1} min^{-1}

As 1 MET $= 3.5$ ml kg^{-1} min^{-1} the target training zone of 13.4–16.7 ml kg^{-1} min^{-1} is equal to 4–5 METs.

Exercise intensity in a **resistance-training program** is usually expressed as either a percentage of 1-RM or as the completion of 8–12-RM. The ACSM (2000) recommend that the average healthy individual should complete a minimum of 8–10 exercises involving the major muscle groups, with a minimum of one set of 8–12-RM. Older or deconditioned persons should aim for 10–15 repetions. **Flexibility** exercises should be held to a position of mild discomfort.

Exercise duration

The exercise duration is very much dependent upon the intensity prescribed for the activity. Exercising at a high intensity for long durations will increase the incidence of injury, and is also likely to be intolerable for many people. Multiplying the exercise intensity by the exercise duration provides an index of **exercise volume**. The same exercise volume can be achieved with high-intensity, short-duration activity as with low-intensity, high-duration activity. The ACSM (2000) recommend a duration of 20–60 minutes (not including warm up and cool down) for both **cardiorespiratory** and **resistance** exercise programs. This duration may be achieved through **continuous activity** or via shorter **intermittent** (minimum of 10 minutes) exercise bouts. Once again the practitioner must reflect on the client's initial fitness, and indeed may choose to begin with 5-minute exercise programs for deconditioned persons. Sessions longer than 60 minutes may lead to drop-out from the program. **Flexibility** exercises should be held for 10–30 seconds for static stretches, and each stretch should be repeated 3–4 times.

Exercise frequency

The number of exercise sessions conducted per week, or the exercise frequency, is again dependent upon the initial fitness of the client. Deconditioned persons may benefit from one to two sessions per week in the first instance. Optimal benefits are obtained from exercising between 3–5 days per week. The practitioner must decide upon the frequency based upon balancing the caloric goal and the likelihood of injury/loss of adherence. The ACSM (2000) recommend that an exercise prescription targets 150–400 kcal of energy expenditure per day. Sections A and J provide more details on kilocaloric expenditure. **Flexibility** exercises should be conducted 2–3 days per week, and can be included in the cool down component of the training session.

Rate of progression

The rate of progression through an exercise program (*Fig. 1*) depends upon the client's goals, age, current health status and initial fitness level. The ACSM (2000) recommend three stages: (1) the initial stage; (2) the improvement stage; and (3) the maintenance stage. The **initial stage** should last 4 weeks and should involve light activities that minimize muscle soreness and injury. This stage will not take the exercise intensity above 60% HRR, or the duration above 30 minutes. The **improvement stage** lasts 4–5 months and allows for the gradual increase of exercise intensity and duration to reach the required goals. The exercise duration and intensity should never be increased during the same exercise session and the rate of change should be slower for older and deconditioned persons. The **maintenance stage** follows the improvement stage and lasts, in an ideal world, for ever. The most important aspects of this stage are to make exercise enjoyable and to reassess goals and physical fitness at regular intervals. The aim of this stage is to maintain a reasonable level of fitness and to ensure a continuation of habitual physical activity.

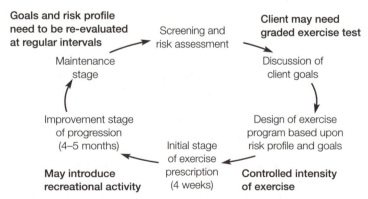

Fig. 1.   Progression through the exercise presciption.

L3 SPECIAL POPULATIONS

Key Notes

Overview

The guidelines reviewed in Section L2 are the general principles of putting together an exercise prescription. These principles should be tailored to suit the client being assessed.

Cardiac patients

Cardiac patients can benefit hugely from exercise. Cardiac rehabilitation can occur at different points following any event or surgery, and guidelines for each of these phases differ accordingly.

Pregnant women

Pregnant women who are healthy can gain the same cardiovascular and musculoskeletal benefits as a non-pregnant woman. Guidelines exist for exercising during pregnancy, and postpartum, with the most important factors relating to safety, thermal stress and energy balance.

Elderly

The benefits of exercise for the elderly are huge. The guidelines for exercise prescription for the elderly are very similar to the generic guidelines but with emphasis on safety and slow progression.

Related topics

Exercise and disease (K3)

Guidelines for exercise prescription (L2)

Overview

The guidelines stated above provide a guide to producing an exercise prescription for any population. The art of prescription is to tailor the exercise program to the client, based upon the screening of the client. Thus deconditioned persons might start exercising at 45% HRR, once per week for 5 minutes per session. Additionally, clients might present as cardiac, stroke, Parkinson's, cerebral palsy, diabetic or arthritic patients (to name but a few), or indeed may be elderly or pregnant. It is beyond the scope of this chapter to review the guidelines for every potential eventuality; however the general guidelines are adapted in these cases to match the requirements and contraindications for these client groups. A few examples are provided below, however further details and guidelines for exercise programs for a number of special populations can be found in ACSM (2000), ACSM (2002) and ACSM (1994).

Cardiac patients

Traditionally **cardiac rehabilitation** programs were categorized into four phases: (I) inpatient; (II) up to 12 weeks of supervised exercise following hospital discharge; (III) a longer-length supervised program with no ECG monitoring; and (IV) a limited supervision program with no ECG monitoring. More recently cardiac rehabilitation has been thought to be best applied to the individual requirements of the patient for their return to health and vocation.

As in the general guidelines cardiac patients need to be screened prior to producing an exercise prescription. This sounds strange given that the patient has a recognized problem or pathology, but screening for the very nature and severity of the pathology is imperative. The **American Association of Cardiovascular and Pulmonary Rehabilitation (AACVPR)** and the **American Heart Association (AHA)** both have risk stratification criteria, and contraindications to exercise for cardiac patients. These criteria can be seen in detail in ACSM (2000) but based on criteria such as ventricular function, functional capacity, angina symptoms and results of a graded exercise test, patients are categorized into low, moderate and high risk of cardiac complications during exercise. The risk aids the practitioner in deciding when and how to begin the exercise prescription, but should be treated only as a part of the overall assessment of requirements.

The aim of the immediate (**within 48 hours**) post-operative or post-event prescription is to expose the patient to changes in orthostatic and gravitational stress. This can be undertaken with stretching and movement of limbs, and with changes in posture via intermittent sitting and standing. This period of time is also ideal for patient **counselling** and **education**.

The aim of **outpatient rehabilitation** is generally to monitor the clinical status of the patient, modify the risk factors associated with the pertinent disease, improve the aerobic tolerance of the patient and to improve general quality of life. Where possible patients should be encouraged to participate in a wide range of activities in order to promote total physical conditioning.

For cardiorespiratory exercise training it is important that the **exercise intensity** is set at a limit that evokes a training effect, but does not pass above the limit that evokes abnormal clinical signs and symptoms. For most deconditioned cardiac patients this intensity is between 40–50% of $\dot{V}O_2R$ and HRR (see Section L2). However, monitoring heart rate does not always provide the best index of ischemia or arrythmia during exercise. The rating of perceived exertion offers a good adjunct to heart rate monitoring, especially if patients are on β-**blockers**. Early phase training should aim for 11–13, or 'fairly light' to 'somewhat hard' on the 6–20 Borg scale, whilst phase III and IV training can increase to 12–15. For some patients knowledge of when **myocardial ischemia** occurs during exercise gained from exercise or stress testing is imperative for setting the exercise prescription intensity. A peak exercise heart rate of 10 beats min^{-1} below this threshold is advised by the ACSM (2000).

The **frequency** of exercise training should be limited to twice per week in phase II rehabilitation but increased to 3–5 times as the prescription progresses, whilst **duration** should be tailored to the patient throughout the program. The **progression** through the program should concentrate on achieving 20–30 minutes continuous exercise before increasing intensities, but again this is dependent upon the functional capacity of the patient. Before progressing onto an unsupervised program the patient should have a functional capacity of 8 METs or more, have stable blood pressure and heart rate at rest, have no cardiac symptoms at rest and should have normal blood pressure responses but no abnormal ECG responses to exercise.

Low- to moderate-risk cardiac patients, especially those whose jobs entail upper body work, can benefit from **resistance training**. The contraindications to resistance training are similar to those for cardiorespiratory work, with the additional criteria of needing a functional capacity of greater than 5 METs without ischemic or anginal symptoms.

Pregnant women   Exercise during **pregnancy** may have implications for the mother and the fetus that may continue into the postpartum period. Most of these implications are positive and there are few instances where exercise should be precluded for healthy, pregnant women. The **American College of Obstetricians and Gynecologists** (ACOG) has published guidelines for exercise during pregnancy and the postpartum period (ACOG, 2003). These guidelines provide full information on the contraindications and recommendations for exercise. The **absolute contraindications** to exercise mainly revolve around cardiovascular and pulmonary disease, and history of complications during pregnancy. The **relative contraindications** include criteria related to obesity, diabetes, heavy smoking, hypertension, anemia, maternal cardiac arrhythmia, thyroid disease, bronchitis and a severely sedentary lifestyle. In the absence of contraindications pregnant women are recommended to engage in regular, moderate-intensity physical activity.

The exercise prescription for pregnant women should include the same elements as for any health-related exercise program. **Aerobic exercise** should include rhythmic large muscle group activities such as jogging, walking, cycling, swimming, etc., although SCUBA-diving and supine activities should be avoided. Common sense would perhaps suggest keeping away from contact sports, and sports with a high risk of injury or falling. The **intensity** for aerobic exercise should not really need to differ from the generic exercise guidelines, 60–70% of maximal heart rate is recommended for previously sedentary pregnant women whilst 70–80% is recommended for previously active women. However, because of the large variability in maternal heart rate responses to exercise, target heart rates should be avoided and replaced with **RPE** as a monitoring tool. Recommendations are 12–14 (somewhat hard) on the Borg scale.

The **duration** of exercise recommended for pregnant women is dictated by concerns for regulating **thermal stress** and **energy balance**. During pregnancy heat production is increased above non-pregnant levels, and this increase continues during exercise. If heat production during exercise exceeds the ability to dissipate the heat, core body temperature will rise and threaten the health of the fetus. Moderate-intensity activities in a thermoneutral environment appear to increase maternal body temperature by less than 1.5°C over 30 minutes. This is a safe limit for pregnant women. After the 13th week of pregnancy an extra 300 kcal of energy per day are required to meet the metabolic demands of pregnancy. Obviously, the additional demands of exercise should be monitored such that the mother consumes enough energy to balance that utilized in pregnancy and exercise. Many studies have indicated that a negative energy balance during pregnancy leads to fetal growth restriction and a consequent low birth weight. This is not the case for women who exercise during pregnancy but ensure that energy balance is maintained. In order to restrict large increases in thermal stress and kilocaloric expenditure in pregnant women, activities of longer than 45 minutes should not be recommended, and accumulating short bouts of exercise seems appropriate. Accumulating 30 minutes on each day of the week is advised for optimal health benefits and **progression** should involve reaching this level.

Resistance training during pregnancy is recommended to aid general musculoskeletal function. The training should include low weights and high repetitions (12–15) for all major muscle groups, and should avoid isometric contractions and heavy weight-lifting. As a rule it is generally accepted that pregnant women who have not participated in resistance training prior to pregnancy should not begin whilst pregnant. **Flexibility training** during pregnancy is not recommended because of the risk of hypermobility.

Postpartum training should progress slowly from a decreased level than during pregnancy, and should only begin after careful consideration by the physician. Exercise for pregnant women with relative contraindications should be fully monitored by a clinician.

Elderly

The **benefits** of exercise for the aging population are enormous. Not only does exercise increase cardiovascular and muscular fitness, it decreases the risk of certain diseases, it aids in the rehabilitation from surgery, it decreases the risk of falling and it helps in the maintenance of independent living. Any individual over the age of 40 years for men, and 50 years for women require **medical clearance** prior to participation in an exercise program. Once medical clearance is gained the exercise program should hinge upon safety, and this might mean using chairs as supports, and if music is used a low-tempo variety. The prescription is conducted as in the generic guidelines with a full baseline assessment undertaken. Prediction of maximal **functional capacity** can be undertaken using protocols designed specifically for elderly groups (for example the Naughton protocol). If a treadmill is used for this purpose handrails should be provided for safety, indeed a cycle ergometer may provide the safer option.

The elderly should aim to accumulate at least 30 minutes of moderate exercise on most days of the week. The **intensity** guidelines for the generic guidelines apply to the elderly, although progression and short bouts in accumulation may be better suited. In order to decrease the risk of musculoskeletal injury and of drop out, longer duration **recreational activities** such as walking or dancing are recommended. **Resistance training** protects against the loss of lean tissue associated with the aging process. The ACSM (2000) recommend that elderly participants perform at least one set of 8–10 exercises of all the major muscle groups using 10–15 repetitions, or an RPE rating of 12–13. This should occur for 20–30 minutes, twice a week, with at least 48 hours of rest between each session. The exercises should be performed slowly through the full range of motion, should be fully supervised and should progress by increasing the number of repetitions before increasing the resistance. Many elderly groups successfully induce resistance-training responses using resistance bands, as opposed to weights. **Flexibility** exercises should aim at including every major joint, and should be static 10–30-second stretches held at mild discomfort, not causing pain. Four repetitions per stretch, 2–3 times per week are recommended, always following a warm up, and never including ballistic stretches.

FURTHER READING

Section A

Anderson, G.S. and Rhodes, E.C. (1989) A review of blood lactate and ventilatory methods of detecting transition thresholds. *Sports Medicine* 8(1): 43–55.

Brooks, G.A. (1991) Current concepts in lactate exchange. *Medicine and Science in Sports and Exercise* 23: 895–906.

Gaesser, G.A. and Brooks, G.A. (1975) Muscular efficiency during steady-rate exercise: effects of speed and work rate. *Journal of Applied Physiology* 38(6): 1132–1139.

Myers, J. and Bellin, D. (2000) Ramp exercise protocols for clinical and cardiopulmonary exercise testing. *Sports Medicine* 30(1): 23–29.

Sutton, J.R. (1992) VO_{2max} – new concepts on an old theme. *Medicine and Science in Sports and Exercise* 24(1): 26–29.

Svedahl, K. and MacIntosh, B.R. (2003) Anaerobic threshold: the concept and methods of measurement. *Canadian Journal of Applied Physiology* 28(2): 299–323.

Section B

Bronk, R. (1999) *Human Metabolism*. Addison Wesley Longman Ltd, Harlow, Essex, UK.

Hames, B.D., Hooper, N.M. and Houghton, J.D. (1997) *Instant Notes in Biochemistry*. Bios Scientific, Oxford.

Houston, M.E. (2001) *Biochemistry Primer for Exercise Sciences*. Human Kinetics. Champaign, Il. USA.

Maughan, R.J., Gleeson, M. and Greenhaff, P.L. (1997) *Biochemistry of Exercise and Training*. Oxford University Press, Oxford.

Maughan, R.J. and Gleeson, M (2004) *The Biochemical Basis of Sports Performance*. Oxford University Press, Oxford.

Section C

Brooks, S.V. (2003) Current topics for teaching skeletal muscle physiology. *Advances in Physiology Education* 27(4): 171–182.

Castro, M.J., McCann, D.J., Shaffrath, J.D. and Adams, W.C. (1995) Peak torque per unit cross-sectional area differs between strength trained and untrained young adults. *Medicine and Science in Sports and Exercise* 27(3): 397–403.

D'Antona, G., Pellegrino, M.A., Adami, R., Rossi, R., Carlizzi, C.N., Canepari, M., Saltin, B. and Bottinelli, R. (2003) The effect of ageing and immobilization on structure and function of human skeletal muscle fibres. *Journal of Physiology* 15, 552: 499–511.

Faulkner, J.A. (2003) Terminology for contractions of muscles during shortening, while isometric, and during lengthening. *Journal of Applied Physiology* 95: 455–459.

Jones, D.A. and Round, J.M. (1990) *Skeletal Muscle in Health and Disease*. Manchester University Press, Manchester, UK.

Ross, A. and Leveritt, M. (2001) Long-term metabolic and skeletal muscle adaptations to short-sprint training: implications for sprint training and tapering. *Sports Medicine* 31(15): 1063–1082.

Thompson, L.V.J. (2002) Skeletal muscle adaptations with age, inactivity, and therapeutic exercise. *Orthopaedics and Sports Physical Therapy* 32(2): 44–57.

Section D

Dempsey, J.A. (1986) Is the lung built for exercise? *Medicine and Science in Sports and Exercise* 18, 143–155.

Dempsey, J.A. and Haskell, W.L. (2004) ACSM, MSSE®, and cardiovascular and respiratory physiology. *Medicine and Science in Sports and Exercise* 36: 2–3.

Demspey, J.A. and Wagner, P.D. (1999) Exercise-induced arterial hypoxemia. *Journal of Applied Physiology* 87: 1997–2006.

Dempsey, J.A., Sheel, A.W., Haverkamp, H.C., Bancock, M.A. and Harms, C.A. (2003) Pulmonary system limitations to exercise in health. *Canadian Journal of Applied Physiology* 28, S2–S24.

Forster, H.V. (2000) Exercise hyperpnea: Where do we go from here? *Exercise and Sport Science Reviews* 28: 133–137.

McClaran, S.R., Babcock, M.A., Pegelow, D.F., Reddan, W.G. and Dempsey, J.A. (1995) Longitudinal effects of aging on lung function at rest and exercise in healthy active fit elderly adults. *Journal of Applied Physiology* 78: 1957–1968.

Olfert, I.M., Balouch, J., Kleinsasser, A., Knapp, A., Wagner, H., Wagner, P.D. and Hopkins, S.R. (2004) Does gender affect pulmonary gas exchange during exercise? *Journal of Physiology* 557: 529–541.

Rundell, K.W. and Jenkinson, D.M. (2002) Exercise-induced bronchospasm in the elite athlete. *Sports Medicine* 32: 583–600.

West, J.B. (2004) Vulnerability of pulmonary capillaries during exercise. *Exercise and Sport Science Reviews* 32, 24–30.

West, J.B. and Wagner, P.D. (1998) Pulmonary gas exchange. *American Journal of Respiratory Critical Care Medicine* 157: S82–S87.

Section E

Achten, J. and Jeukendrup, A.E. (2003) Heart rate monitoring: applications and limitations. *Sports Medicine* 33: 517–538.

Bowles, D.K., Woodman, C.R. and Laughlin, M.H. (2000) Coronary smooth muscle and endothelial adaptations to exercise training. *Exercise and Sport Science Reviews* 28: 57–62.

Carter, J.B., Banister, E.W. and Blaber, A.P. (2003). Effect of endurance exercise on autonomic control of heart rate. *Sports Medicine* 33: 33–46.

Coyle, E.F. and González-Alonso, J. (2001). Cardiovascular drift during prolonged exercise: New perspectives. *Exercise and Sport Science Reviews* 29: 88–92.

Seals, D.R. (2003). Habitual exercise and the age-associated decline in large artery compliance. *Exercise and Sport Science Reviews* **31**: 68–72.

Sharma, S., Whyte, G., Elliot, P., Padula, M., Kaushal, R., Mahon, N. and McKenna, W.J. (1999) Electrocardiographic changes in 1000 highly trained junior elite athletes. *British Journal of Sports Medicine* **33**: 319–324.

Spina, R.J. (1999) Cardiovascular adaptations to endurance exercise training in older men and women. *Exercise and Sport Science Reviews* **27**: 317–332.

Thirup, P. (2003) Hematocrit: within-subject and seasonal variation. *Sports Medicine* **33**: 231–243.

Thompson, P.D. (2004) Historical concepts of the athlete's heart. *Medicine and Science in Sports and Exercise* **36**: 363–370.

Section F

Aagaard, P. (2003) Training-induced changes in neural function. *Exercise and Sport Science Reviews* **31**: 61–67.

Braun, B. and Horton, T. (2001) Endocrine regulation of exercise substrate utilization in women compared to men. *Exercise and Sport Science Reviews* **29**: 149–154.

Carrol, T.J., Riek, S. and Carson, R.G. (2001) Neural adaptations to resistance training: implications for movement control. *Sports Medicine* **31**: 829–840.

Carter, J.B., Banister, E.W. and Blaber, A.P. (2003). Effect of endurance exercise on autonomic control of heart rate. *Sports Medicine* **33**: 33–46.

Godfrey, R.J., Madgwick, Z. and Whyte, G.P. (2003). The exercise-induced growth hormone response in athletes. *Sports Medicine* **33**: 599–613.

Kraemer, W.J., Hakkinen, K., Newton, R.U., Nindl, B.C., Volek, J.S., McCor, M., Gotshalk, L.A., Gordon, S.E., Fleck, S.J., Campbell, W.W., Putukian, M. and Evans, W.J. (1999) Effects of heavy resistance training on hormonal response patterns in younger vs. older men. *Journal of Applied Physiology* **87**: 982–992.

Kraemer, W.J., Fleck, S.J., Maresh, C.M., Ratamess, N.A., Gordon, S.E., Goe, K.L., Harman, E.A., Frykman, P.N., Volek, J.S., Mazzetti, S.A., Fry, A.C., Marchitelli, L.J. and Patton, J.F. (1999) Acute hormonal responses to a single bout of heavy resistance exercise in trained power lifters and untrained men. *Canadian Journal of Applied Physiology* **24**: 524–537.

Smilios, I., Pilianidis, T., Karamouzis, M. and Tokmakidis, S.P. (2003) Hormonal responses after various resistance exercise protocols. *Medicine and Science in Sports and Exercise* **35**: 644–654.

Wideman, L., Weltman, J.Y., Hartman, M.L., Veldhuis, J.D. and Weltman, A. (2002) Growth hormone release during acute and chronic and resistance exercise: recent findings. *Sports Medicine* **32**: 987–1004.

Zhou, S. (2000) Chronic neural adaptations to unilateral exercise: Mechanisms of cross education. *Exercise and Sport Science Reviews* **28**: 177–184.

Section G

Jeukendrup, A.E. (1998) Fat as a fuel during exercise. In: Berning, J.R. and Steen, S.N. (eds) *Nutrition for Sport and Exercise*. Gaithersburg, MD: Aspen Publishers.

Karlsson, J. (1997) *Antioxidants and Exercise*. Champaign, IL: Human Kinetics.

MacLaren, D.P.M. (2003). Creatine. In: Mottram, D (ed.) *Drugs in Sport*, 3rd edn. London: E&FN Spon.

Manore, M. and Thompson, J. (2000) *Sport Nutrition for Health and Performance*. Champaign, IL: Human Kinetics.

Maughan, R.J. (1999) Nutritional ergogenic aids and exercise performance. *Nutrition Research Reviews* **12**: 255–280.

Section H

American College of Sports Medicine (1998). Position stand on the quantity and quality of exercise for developing and maintaining cardiorespiratory and muscular fitness and flexibility in healthy adults. *Medicine and Science in Sports and Exercise* **30**: 975–991.

Bompa, T.O. (1999) *Periodization Training for Sports*. Human Kinetics: Champaign, Illinois.

Bompa, T.O. and Cornachia, L. (1998) *Serious Strength Training*. Human Kinetics: Champaign, Illinois.

Plowman, S.A. and Smith, D.L. (2002) *Exercise Physiology for Health, Fitness, and Performance*. Benjamin Cummings: San Francisco.

Section I

Atkinson, G. and Reilly T. (1996) Circadian variation in sports performance. *Sports Medicine* **21**: 292–312.

Cheuvront, S.N. and Haymes, E.M. (2001) Thermoregulation and marathon running: biological and environmental influences. *Sports Medicine* **31**: 743–762.

Coris, E.E., Ramirez, A.M. and Van Durme, D.J. (2004) Heat illness in athletes: the dangerous combination of heat, humidity and exercise. *Sports Medicine* **34**: 9–16.

Febbraio, M.A. (2001) Alterations in energy metabolism during exercise and heat stress. *Sports Medicine* **31**: 47–59.

Ferretti, G. (2001) Extreme human breath-hold diving. *European Journal of Applied Physiology* **84**: 254–271.

Fulco, C.S., Rock, P.B. and Cymerman, A. (1998) Maximal and submaximal exercise performance at altitude. *Aviation and Space Environmental Medicine* **69**: 793–801.

Hoppeler, H and Vogt, M. (2001) Hypoxia training for sea-level performance. Training high–live low. *Advances in Experimental Medicine and Biology* **502**: 61–73.

Noakes, T.D. (2000) Exercise and the cold. *Ergonomics* **43**: 1461–1479.

Shirreffs, S.M. and Maughan, R.J. (2000). Rehydration and recovery of fluid balance after exercise. *Exercise and Sport Science Reviews* **28**: 27–32.

Smolander, J. (2002) Effect of cold exposure on older humans. *International Journal of Sports Medicine* **23**: 86–92.

Strauss, M.B. and Borer, R.C., Jr. (2001) Diving medicine: contemporary topics and their controversies. *American Journal of Emergency Medicine* **19**: 232–238.

Waterhouse, J., Reilly, T. and Atkinson, G. (1997) Jet lag. *Lancet* **350**: 1611–1616.

Wilber, R.L. (2001) Current trends in altitude training. *Sports Medicine* **31**: 249–265.

Section J

Bouchard, C. (2000) *Physical Activity and Obesity*. Human Kinetics

Eston, R. and Reilly, T. (2001) *Kinanthropometry and exercise physiology laboratory manual: Vol. 1: Anthropometry: tests, procedures and data.* Routledge.

Manore, M. and Thompson, J. (2000) *Sport Nutrition for Health and Performance.* Human Kinetics.

Montoye, H.J. (1996) *Measuring physical activity and energy expenditure.* Human Kinetics.

Section K

Ades PA, Green NM, Coello CE. (2003) Effects of exercise and cardiac rehabilitation on cardiovascular outcomes. *Cardiology Clinics* 21(3): 435–48, viii.

American College of Sports Medicine (1994) Exercise for patients with coronary artery disease. Position stand. *Medicine and Science in Sports and Exercise* **26:** i–v.

American College of Sports Medicine (2000) *ACSM's Guidelines for Exercise Testing and Prescription* (Sixth Ed), Lipincott Williams and Wilkins, Philadelphia, USA.

American College of Sports Medicine (2001) *ACSM's Resource Manual for Guidelines for Exercise Testing and Prescription* (Fourth Ed), Lipincott Williams and Wilkins, Philadelphia, USA.

American College of Sports Medicine (2002) *ACSM's Resources for Clinical Exercise Physiology.* Lippincott Williams and Wilkins, Philadelphia, USA.

Blair SN. (2003) Revisiting fitness and fatness as predictors of mortality. *Clinical Journal of Sport Medicine* **13**(5): 319–20.

Bouchard, C., Shephard, R.J. and Stephens, T. (1992) *International Consensus Symposium on Physical Activity, Fitness, and Health* (Toronto, Canada)

Church, T.S., Cheng, Y.J., Earnest, C.P., Barlow, C.E., Gibbons, L.W., Priest, E.L. and Blair, S.N. (2004) Exercise capacity and body composition as predictors of mortality among men with diabetes. *Diabetes Care* **27**(1): 83–88.

Kurl, S., Laukkanen, J.A., Rauramaa, R., Lakka, T.A., Sivenius, J. and Salonen, J.T. (2003) Cardiorespiratory fitness and the risk for stroke in men. *Archives of Internal Medicine* **163**(14): 1682–1688.

Lee, I.M., Sesso, H.D., Oguma, Y. and Paffenbarger, R.S., Jr. (2003) Relative intensity of physical activity and risk of coronary heart disease. *Circulation* **107**(8): 1110–1116.

Shephard, R.J. (1978) *Physical Activity and Aging.* Croom Helm, London.

Wallace, J.P. (2003) Exercise in hypertension. A clinical review. *Sports Medicine* **33**(8): 585–598.

Section L

American College of Sports Medicine Position Stand (1998) The recommended quantity and quality of exercise for developing and maintaining cardiorespiratory and muscular fitness, and flexibility in healthy adults. *Medicine and Science in Sports and Exercise* 30(6): 975–991.

American College of Sports Medicine Position Stand (1998) Exercise and physical activity for older adults. *Medicine and Science in Sports and Exercise* **30**(6): 992–1008.

American College of Sports Medicine (1994) Exercise for patients with coronary artery disease. Position stand. *Medicine and Science in Sports and Exercise* **26:** i–v.

American College of Sports Medicine (2000) *ACSM's Guidelines for Exercise Testing and Prescription,* 6th edn. Lippincott, Williams and Wilkins: Philadelphia, PA.

American College of Sports Medicine (2001) *ACSM's Resource Manual for Guidelines for Exercise Testing and Prescription,* 4th edn. Lippincott, Williams and Wilkins, Philadelphia, PA.

American College of Sports Medicine (2002) *ACSM's Resources for Clinical Exercise Physiology.* Lippincott, Williams and Wilkins, Philadelphia, PA.

Artal, R. and O'Toole, M. (2003) Guidelines of the American College of Obstetricians and Gynecologists for exercise during pregnancy and the postpartum period. *British Journal of Sports Medicine* 37: 6–12.

Bouchard, C., Shephard, R.J. and Stephens, T. (1992) *International Consensus Symposium on Physical Activity, Fitness, and Health.* Toronto, Canada.

Dionne, I.J., Ades, P.A. and Poehlman, E.T. (2003) Impact of cardiovascular fitness and physical activity level on health outcomes in older persons. *Mechanisms of Ageing and Development* **124**(3): 259–267.

Heyward, V.H. (1997) *Advanced Fitness Assessment and Exercise Prescription,* 3rd edn. Human Kinetics, Champaign, IL.

Swain, D.P. (2000) Energy cost calculations for exercise prescription: an update. *Sports Medicine* 30(1): 17–22.

Welsch, M.A., Pollock, M.L., Brechue, W.F. and Graves, J.E. (1994) Using the exercise test to develop the exercise prescription in health and disease. *Primary Care* **21**(3): 589–609.

INDEX